THE
BEAUTY
OF
LIGHT

BEN BOVA

THE BEAUTY OF LIGHT

WILEY

Wiley Science Editions

JOHN WILEY & SONS, INC.

New York · Chichester · Brisbane · Toronto · Singapore

Publisher: Stephen Kippur
Editor: David Sobel
Managing Editor: Ruth Greif
Editing, Design, and Production: G&H SOHO, Ltd.
Photo Research: Corinne McCormick, John L. Raybin, John Schultz

Library of Congress Cataloging-in-Publication Data

Bova, Ben, 1932-
    The beauty of light / Ben Bova.

        p. cm.—(Wiley science editions)
    Bibliography: p. 341
    ISBN 0-471-62580-9
    1. Light—Popular works.   I. Title.   II. Series.
QC358.5.B68 1988
635—dc19   88-12228
                    CIP

Printed in the United States of America

88 89 10 9 8 7 6 5 4 3 2 1

To Sarah King and Donald P. Cushman,
who have greatly helped to enlighten me.

Illustration Credits

Page 92: Fritz Prenzel/Animals Animals.

Page 93: From *Human Biology in Health and Disease,* by Shirley R. Burke/ John Wiley & Sons, Inc./© 1975.

Page 100: © Robert P. Comport.

Page 103: AIP Niels Bohr Library/Physics Today Collection.

Page 106: PAR/NYC Archive from "A Pictorial History of the World's Great Nations," by Charlotte Yonge 1882.

Page 109: From *Light and Color in Nature and Art,* by Samuel Williamson/ John Wiley & Sons, Inc./© 1983.

Page 115: Courtesy: Biblioteca Nazionale Centrale, Florence.

Page 120: AIP Niels Bohr Library.

Page 131: From *Light and Color in Nature and Art,* by Samuel Williamson/ John Wiley & Sons, Inc./©1983.

Page 137: AIP Niels Bohr Library.

Page 143: *The New York Times.*

Page 147: United States Air Force.

Page 151 (top): Art Resource Center.

Page 151 (bottom): AIP Meggers Gallery of Nobel Laureates.

Page 155: The American Museum of Natural History.

Page 161: Woodcut by Durer from the 1525 edition of his textbook on perspective and proportion.

Page 163: Art Resource.

Page 165: From *Light and Color,* by R. Daniel Overheim/John Wiley & Sons, Inc./©1982.

Page 171: William Vandivert.

Page 172: Alistair B. Fraser.

Page 173: From Street-Crestalt Completion Test, 1931, published by Columbia University Teacher's College.

Page 179: Richard Hutchings/Photo Researchers.

Page 182: From *Light and Color in Nature and Art,* by Samuel Williamson/John Wiley & Sons, Inc./© 1983.

Page 185: From *Light and Color in Nature and Art,* by Samuel Williamson/John Wiley & Sons, Inc./© 1983.

Page 188: From *Light and Color in Nature and Art,* by Samuel Williamson/John Wiley & Sons, Inc./© 1983.

Page 191: 1975 by AURA Inc., Kitt Peak National Observatory.

Page 193: International Museum of Photography at George Eastman House.

Page 196: PAR/NYC Archive.

Page 197: Courtesy Eastman Kodak Co.

Page 204: Edison Natural History Society.

Page 208: French Government Tourist Office.

Page 210 (top): Folger Shakespeare Library Prints.
Page 214: PAR/NYC Archive.
Page 221: Eastman Kodak Co.
Page 225: Dr. Theodore Maiman.
Page 236: Laboratory for Laser Energetics.
Page 244 (top): NASA.
Page 244 (bottom): Photo Researchers.
Page 248: American Bank Note Holographics, Inc.
Page 250: Hologram produced by Mr. Rainbows, P.O. Box 27056, Philadelphia, PA 19118.
Page 252: A,T&T Bell Laboratories.
Page 256: United States Army Strategic Defense Command and TRW Defense Systems Group.
Page 258: Lawrence Livermore Laboratories.
Page 263: NASA/Marshall Space Flight Center.
Page 269: A,T&T Bell Laboratories
Page 274: Texas Instruments.
Page 277: NASA.
Page 279: PAR/NYC Archive.
Page 281: Tom Kelly from PAR/NYC Archive.
Page 287: National Optical Astronomy Observatories.
Page 290: Don Yeomans.
Page 291: Tom Howarth.
Page 295: American Museum of Natural History.
Page 297: American Museum of Natural History.
Page 300: Kitt Peak National Observatory.
Page 303: Mt. Wilson and Palomar Observatories.
Page 315: NASA.
Page 316: CSIRO Radio Physics Division Photo Lab.
Page 319: Sacramento Peak Observatory.
Page 325: Astronomical Society of the Pacific.
Page 328: © Helmut Wimmer.
Page 332: NASA.
Page 333: Hale Observatories.
Page 335: Hale Observatories.

Color photos following page 78:

Plate 1: Sinclair Stammers/Science Photo Library/Photo Researchers.
Plate 2: NASA.
Plate 3: Animals Animals.
Plate 4: Lennart Nilson.

Plate 5: Lander Photos/NASA.
Plate 6: Nicholas Foster/The Image Bank.
Plate 7: F. Whitney/The Image Bank.
Plate 8: NASA.
Plate 9: Bausch & Lomb.

Color photos following page 174:

Plate 10: Art Resource.
Plate 12: The Corning Museum of Glass.
Plate 14: Art Resource.
Plate 15: Art Resource.

Color photos following page 238:

Plate 19: Art Resource.
Plate 21: Laboratory for Laser Energetics.
Plate 22: PAR/NYC.
Plate 23: Art Resource.
Plate 25: Astronomical Society of the Pacific.
Plate 26: Astronomical Society of the Pacific.
Plate 27: National Optical Astronomy Observatories.

Author's Foreword

THIS BOOK began with the title *The Beauty of Light* and the idea that I wanted to write about how light affects us and how we use light in art, science, industry, and our everyday lives.

The more I studied and wrote on the subject, the more I found to study and write about. Light affects us in so many ways. We use light every moment of our waking lives, and we even use the visual cortex back there in the rear hemisphere of the brain while we sleep.

As a result, this book touches on hundreds of topics—lightly. What I have tried to do is to simply sit down and chat with you, to tell you about all the fascinating, surprising, and wonderful things that we do with light and that light does for us. This is by no means a textbook. It does not pretend to offer the complete or definitive story on any of the topics it discusses. It is meant merely to inform you, perhaps to delight you, and to intrigue you enough so that you will understand and appreciate the wonders that strike your eye every day.

The White Queen in Lewis Carroll's *Through the Looking Glass* exclaimed, "Why, sometimes I've believed as many as six impossible things before breakfast!" This book will not deal with impossible things, but I fondly hope that you will find surprising things, new ideas, and fresh insights on just about every page. I have tried to keep the tone of the writing informal and conversational. Still, I believe that you will learn a good deal from these pages. At the very least, you ought to get a few smart new items of information to spring on your friends at your next party.

The plan of the book is straightforward. In the first section we examine how light has affected life from the beginnings of the Earth down to the present day; we pay particular attention, of course, to light's effects on human life. Then we look at how scientists have tracked down the very nature of light and learned what it is and how it works. In the third section we see how we use light in art, science, and technology—from Ice Age cave paintings to modern motion pictures and rock concerts. Finally we turn our vision toward the stars to show how those lights in the sky are telling us the story of creation and how astronomers are trying to decipher their messages.

Each section can be read independently of the others. If you are not terribly interested in what the physicists have learned about the nature of light, you can skip the second section, for example, and go on to see how artists and engineers and theater folk use light.

Some of these topics are discussed in greater detail than others. I must confess that the decisions on how deeply to delve into one subject or another were based on my personal preferences. I concentrated on those areas that I believe are most fascinating and spent less time on those that seem more commonplace, to me. I hope you share my interests. If not, rest assured that there are other books in which you can read all the details you want about any subject covered here.

A book of this breadth cannot be produced by one person alone. Many people—friends, colleagues, scientists, and artists—have shared their thoughts and their time with me. I am particularly indebted to Sarah King, chairman of the communications department of Central Connecticut State University; Mary Beth Lucas, whose intelligence and energetic research activities I especially appreciate (among other things, for finding the size of a blue whale's eye and the diameter of an optical fiber); to Kenneth Jon Rose of the New York University biology department; and to Barbara Sullivan, Elenore Zavez, and Joan Packer of the Central Connecticut State University library. Without their help this book would either be only a pale shadow of what it is now or would never have been completed at all.

I want to express my admiration and gratitude to David Sobel, my editor, who shares my enthusiasm for books about science that are written for the general reader. Most of all, I thank my agent and wife, Barbara Bova, for her love, her support, her patience with my Mediterranean temperament—and for originating both the idea and the title of this book.

Ben Bova
West Hartford, Connecticut

Contents

Full many a glorious morn have I seen
Flatter the mountain-tops with sovereign eye,
Kissing with golden face the meadows green,
Gilding pale streams with heavenly alchemy . . .

William Shakespeare
Sonnet XXXIII

I.

TO

SEE

The best thing that we're put here for's to see . . .

—*Robert Frost*
"The Star-Splitter"

1

Creatures of the Light

"LET THERE BE LIGHT!"

With those words God began the creation of the universe, according to the first chapter of Genesis. The fundamental act of creation was to divide light from darkness.

We are creatures of the light. The light of the Sun gives life to our world. The light of fire is the basis of all human civilization. The power and beauty of light are so fundamental to our lives that we incorporate the metaphors of vision and light into our everyday language without even thinking about it.

I see what you mean, we say, because seeing is believing, isn't it?

Look at it this way, we tell a friend who doesn't quite see things the way we do. Perhaps we can shed more light on the subject if we look deeper into it.

A bright person gets ahead in the world, while a dull one has a more difficult time. Still, it is better to light a candle than to curse the darkness.

Looking at it in a different light, a sunny disposition is much to be preferred over a dark one. And what can make you glow more brightly than seeing the light of your life smiling at you?

Seeing is believing, and seeing is so important to us that we use dozens of different words to describe it. We not only see, we look,

Creation still goes on. New stars are being formed out of dark clumps of interstellar gas and dust in regions such as the Cone Nebula.

watch, behold, distinguish, notice, note, mark, observe, examine, inspect, scan, scrutinize, probe, study, appraise, gaze, glare, peek, peer, and stare. We give something the eye, keep our eye on it, hold it in view. We ascertain, envision, and visualize. And when sudden danger confronts us we shout, "Look out!"

We picture things in our imaginations and sometimes make pictures out of what we have seen with our inner eye. Often the pictures are quite graphic; sometimes they're picturesque.

When we first understand some new fact we say that the dawn has risen or the light has finally shone. We can light a fire under someone who is sluggardly or kindle a flame in the heart of a loved one. Light can mean happy or trivial, easy or simple, fast or loose; it never means dark, brooding, cumbersome, massive, ponderous, unwieldy, or weighty.

Good guys wear white hats; bad guys wear black.

Light is the difference between life and death. It has shaped our world and shaped us. It is the most important medium of information that we have.

Try this experiment: The next time you are at a party, close your eyes and try to identify the people around you. You will hear the voices of those closest to you, but those just a few feet away will merge into a background noise. Compare this to *seeing* the people in the room; there is a world of difference. You can see

4

the spatial relationships among the people in the room: who is standing close to you, who is farther away; who is tall and who is short; who is slim and attractive and who should start a diet. Try getting all that information through your ears or by touch or taste or smell. Not only do you run the risk of being slapped, pummeled, and thrown out of the party; even if the other guests allowed you to pet, sniff, and nip at them, it would take *ages* to gather the information that a single glance would give.

Most of the information we get about the world around us comes into our brain through our eyes. Vision provides more than ten times as much information as hearing, principally because light waves can carry enormously more "bits" of information than sound waves. Our minds depend on sight. Blindness has always been a particular horror to the sighted and the most desperate of handicaps to those who lose their precious sight.

The great scientist Galileo, the man who first turned a telescope to the heavens only to be persecuted and imprisoned by the Church in his old age, wrote in 1638:

> Alas . . . your devoted friend and servant, has been for a month totally and incurably blind; so that this heaven, this earth, this universe, which by my remarkable observations and clear demonstrations I have enlarged a hundred, nay, a thousand fold . . . are now shrivelled up for me into such a narrow compass as is filled by my own bodily sensations.

He was visited in Italy by a young English poet, John Milton, who would one day write *Paradise Lost*. Ironically, tragically, Milton himself went blind some thirty years later, writing:

> When I consider how my light is spent
> Ere half my days in this dark world and wide,
> And that one Talent which is death to hide
> Lodged with me useless . . .

Vision is a gift; sight is the most important of the five senses for most humans.

You can read silently much faster than you can read aloud. Another experiment to try: Time yourself as you read this page to yourself, then time how long it takes to read it aloud to a friend.

If you begin to approach the speed of silent reading, your listener will complain that you are going too fast to be understood. We acquire much larger amounts of information through vision than through hearing.

Light is basic to our moods, our outlooks, our very lives.

When I was a young reporter just out of journalism school, my first newspaper job made me "see the light" of that fact. Among the chores foisted onto new reporters was the task of writing obituaries. It did not take me long to realize that more people died after a spell of dark, gloomy weather than did during fine, sunny weather. While a tiny handful of these deaths were accidents due to slippery roads, most of them were among the elderly or bedridden ill, who simply seemed to lose the will to live when the weather turned dark for days on end.

Take a look at a map of Europe. (Another metaphor of vision!) The rate of suicides per capita, and of alcoholism, runs a fairly strong parallel with the number of cloudy days per year. In cold northern nations, where the sun is seen only briefly during long dark winters, suicides and alcoholism are relatively high. In the sunny Mediterranean climes, the rates are much lower.

Light has shaped our view of the world, our ideas of good and evil, even our religions.

Thy dawning is beautiful in the horizon of the sky,
O living Aton, beginning of life.
When thou risest in the eastern horizon,
Thou fillest every land with thy beauty.

Such was the prayer of the Pharaoh Akhnaton, thirteen centuries before Christ. Ancient man realized that the blinding light of the Sun was the source of life. They associated that overwhelming brightness with all the good things of the world. They worshiped the Sun.

By the sixth century B.C. the Iranian holy man Zoroaster founded a religion based on the light—and the dark. Known to the later Greeks as Zarathushtra, Zoroaster's religion divided the world into light and darkness, truth and falsehood, good and evil. Every man faced the choice of following Ahura Mazda, the lord of light and goodness, or Angra Mainyu, the destructive spirit.

Ancient man realized that the light of the Sun was the source of life. They worshipped the Sun.

Deep in the rugged mountains of eastern Iran, where the dusty hills of Khorasan rise toward the forbidding peaks of Afghanistan, Zoroaster and his followers worshiped not the Sun but *fire*, the light created by man. Ceremonial bonfires were their temples, blazing in the mountain winds, lighting the darkness of those cold ancient nights.

From those fire-lit mountain fastnesses the idea of light as a representation of good and darkness as an avatar of evil spread across the ancient world, carried by the swords of Cyrus, Xerxes, Darius, and the other great conquerors and rulers who established the ancient Persian empire.

Between the fifth and third centuries B.C. the ancient Greeks fought off the Persian armies and eventually, under Alexander the Great, conquered the conquerors. In their centuries-long struggle they came into intimate contact with the religion of Zoroaster. To the Greek philosophers the duality that is at the heart of Zoroastrianism was a new concept; the gods and goddesses of Mount Olympus recognized no such division between good and evil.

Hebrew prophets freed from Babylonian captivity in 536 B.C. by the Persians also were influenced by the dualism of Zoroaster. The concept of an ongoing struggle between good and evil, between light and darkness, became a part of the Hebrew view of the universe. In turn, Jewish thought powerfully influenced the development of Christianity, and Zoroaster's vision of the eternal battle between a creative spirit and a destructive one was incorporated into the Christian theology.

But long before Zoroaster, before Akhnaton, before the invention of writing or the discovery of agriculture or the birth of the first human being, the beauty of light shaped our world and began the long, long process that led to life, to vision, to the human race.

2

From Darkness to Light

IN THE BEGINNING, according to Genesis, "the earth was without form, and void; and darkness was upon the face of the deep."

There are remarkable parallels between the majestic poetry of Genesis and the findings of modern astronomers. In the beginning, our planet Earth was indeed dark and without form.

The Solar System originated roughly 4.6 billion years ago. Four point six thousand million years. Forty-six million centuries ago. The evidence of this enormous age comes from studies of the rocks of the Earth and the Moon, meteorites that fall to Earth from outer space, and the composition and behavior of the Sun.

In that beginning, there was no Sun and no Earth. No planets or moons. Just a vast cloud of dark gas laced with tiny grains of dust, a cloud so huge that it would take light a whole year to cross its span.

Picture the scene as it must have been 46 million centuries ago. There were stars shining in the eternal darkness of space. The Milky Way, that titanic pinwheel colony of billions of stars, glowed steadfastly against the black void even then. But where our Solar System was to be born there was only this dark swirling cloud of dust-laden gas, about a light-year wide.

"The earth was without form, and void, and darkness was upon the face of the deep." How the primeval Earth might have looked, before sunlight penetrated to the surface and life began to reshape our planet.

The cloud consisted of cool gases laced with dark flecks of microscopic dust grains. Astronomers have seen such clouds in interstellar space and are convinced that they are in the process of forming new stars and planetary systems.

Our dust cloud was spinning, swirling in the emptiness of space. As it spun it began to collapse in on itself. The two phenomena are part of the same process: the faster the spin, the faster the collapse; the more the cloud shrank in on itself, the faster it spun. Think of a skater twirling on the ice, arms extended. As she pulls her arms closer to her body, she spins faster and faster until she is a blur of motion. She extends her arms and her spinning slows immediately. Physicists call this the *conservation of angular momentum*. It works for skaters, for atoms, for planets and stars, for whole galaxies.

Most of the gas in the original cloud was hydrogen and helium, while the dust particles contained other elements, such as silicates and metals. As the gas and dust particles sank to the center of the cloud, its core grew into a massive sphere and began to glow. At first its light was sullen and red, caused by the heat of countless megatons of in-falling dust and gas. But deep within the heart of the sphere, unbelievable pressures were building and the temperature rose to thousands, and then millions, of degrees. Finally the temperature and pressure at the center of the sphere became so great that the nuclei of the tortured hydrogen atoms

To the astronomers, light is not only the messenger carrying information from the stars; it is a yardstick. Astronomers measure the incredible distances of space by using the speed of light as a benchmark.

Nothing in the universe travels faster than light. Albert Einstein postulated this in his Special Theory of Relativity in 1905, and the subtlest and most cunning experiments devised by physicists ever since have failed to prove otherwise. Light is the ultimate speedster. In the vacuum of space it travels at a velocity of approximately 186,000 miles per second, the universe's quintessential speed limit. This has given rise to a T-shirt motto worn by cognoscenti at places such as MIT and Caltech:

186,000 MILES PER SECOND
IT'S NOT JUST A GOOD IDEA
IT'S THE LAW!

There are about 31.5 million seconds in a year, which means that light can cross the phenomenal distance of 5.859 trillion miles in a year. That distance, roughly 6 trillion miles, is called a *light-year* by astronomers.

(In metric units the speed of light is just about 300,000 kilometers per second, and a light-year equals 9.45 trillion kilometers.)

Using light-years is much easier on astronomers' nerves than trying to measure the mind-boggling distances in space in terms of miles or kilometers. The Milky Way galaxy, for example, is about 100,000 light-years in diameter. Our Solar System is roughly 30,000 light-years from the galaxy's center.

there began to fuse together, forced into unions that created helium and released the energy we call light.

A star was born. The raging hell of thermonuclear hydrogen fusion at the core of our particular star turned its surface into a blindingly brilliant sphere of light. The Sun began to shine, more than 4 billion years ago.

The battered face of the Moon shows the scars left by the violent early stages of the Solar System's formation. The largest craters were formed no later than 3 billion years ago.

All around the newborn Sun bits and pieces of the dust cloud had been coalescing into larger bodies, pulled together by gravity, colliding, often breaking apart only to come together again. Microscopic grains grew to the size of pebbles, then boulders, mountains, worldlets. The planets were formed in this often-violent process of accretion. We can see the violence that must have marked those early days on the battered face of our own Moon. The giant craters that pockmark the Moon were formed no later than 3 billion years ago, according to the evidence of the rocks brought back to Earth by the Apollo astronauts.

This violent phase of planet building also battered the other worlds of our Solar System. Mars, Mercury, and Venus all bear their scars. So do the moons of distant Jupiter and Saturn. Even on Earth, where wind and weather have erased most of the craters, there are huge structures called astroblemes in Canada, Africa, and elsewhere. Hudson's Bay might be the result of a large meteor impact. The Ashanti Crater in Ghana certainly is. The famous Meteor Crater in Arizona is of much later origin; it was blasted out of the desert only a few thousand years ago by one of the chunks of rock or metal still floating through the Solar System, leftover bits and pieces of the original creation of the worlds. Unless protected, the Arizona crater will be scrubbed away by weathering in a few million years or less.

Some 4 billion years ago the Sun was shining even brighter than it does today and the Earth had formed. But there was darkness on the face of the Earth, and its surface looked more like the pits of hell than the green lovely world we call our home.

The Earth was *hot*. Hot enough to melt rocks. In the process of building this world, dust grains collided and stuck together to form larger particles. These larger particles began to collide, sometimes bouncing away from one another, sometimes sticking together. As the particles swung around in their orbits through the thinning cloud of dust and gas surrounding the newly born Sun, they picked up more dust grains, suffered more collisions with larger particles. Sometimes they broke apart. Sometimes they grew into chunks of rock and metal tens or even hundreds of miles wide. Astronomers called them *planetesimals*.

Eventually—and it may have been as little as only a few thousand years—these growing chunks of rock and metal grew to the sizes of planets such as our Earth, roughly 8,000 miles in diameter. But the energy released by those myriad collisions heated the rocks immensely. And among the elements composing the rocks and metals were radioactive elements such as uranium, thorium, and radium. Our world was nuclear heated from within.

The Earth became molten. The rocks liquefied and flowed just as lava flows from an erupting volcano. The heavier elements sank to the core of our planet, giving our world a dense metallic core surrounded by lighter layers of rock. Gradually the outermost layers of rock radiated their heat away into space, cooled, and the surface of the Earth became solid, trapping the remaining heat beneath the insulating crust of rock.

Some 4 billion years after the formation of the Earth, that primeval heat still seethes beneath our feet. We get glimpses of its fury when a volcano spews molten rock or blasts a mountaintop into powder. Even the "solid ground" of the continents we live on are actually islands of rock floating on the hotter, more plastic rocks beneath them, huge plates of granite and basalt inching this way and that, throwing up mountain chains where they grind slowly into one another, shuddering the ground with earthquakes where they meet.

And deep beneath it all, more than 1,800 miles below the surface, the Earth's core of nickel and iron is still hot enough to be fluid, except at its very heart, where the titanic pressures of a whole world pressing down on it makes the innermost core solid metal.

Even when the surface of the Earth cooled enough to solidify, the planet was still dark and totally unlike the world we know. There was no life. Not a blade of grass or a microscopic bacterium. There were no oceans, no seas. The world was bare rock, covered with thick clouds that must have blanketed the Earth for eons.

Not even the air was the same as we have today. The Earth's atmosphere at that time was composed largely of the gases from the original cloud that had been the progenitor of the Solar System. The atmosphere was mostly hydrogen and helium gas, with smaller amounts of nitrogen and hydrogen compounds such as methane and ammonia.

And water vapor.

That was the most important ingredient of all, even though it was probably the smallest. For our planet Earth is situated at a distance from the Sun where water can exist in all three of its physical forms: as vapor, as ice, and—crucially important for the existence of life—as a liquid.

The black clouds covering the Earth contained a good deal of water vapor, and for untold eons they rained down on the bare rocks below. At first the rocks were still so hot that most of the rain flashed into steam and rose back up through the atmosphere to form clouds again. Gradually, however, as the rocks cooled, liquid water began to collect on the surface of our world. Ponds, lakes, seas, and mighty oceans began to grow.

The clouds eventually broke apart and sunlight smiled down on the surface of a barren, rocky world. But it was a world that was partially covered now with liquid water. It was a world ready to bear life.

Our sister world, Venus, two-thirds closer to the Sun than

we, never developed oceans. It was never cool enough on Venus's surface for liquid water to form. Venus, although just about the same size as Earth, is a barren hellhole of a planet, with surface temperatures hot enough to melt aluminum, an atmosphere of dense choking carbon dioxide, and sulfurous cloud blankets smothering the entire planet.

Sunlight and water. They transformed a barren rocky Earth into a haven of life and, eventually, intelligence.

Sunlight helped to change the Earth's primitive atmosphere. Solar energy heated the atmosphere enough so that the lightest gases, hydrogen and helium, literally boiled away into space. Thanks to ongoing chemical reactions in the atmosphere and on the surface, Earth's atmosphere became a mixture of nitrogen, carbon dioxide, ammonia, methane, a smattering of inert noble gases such as argon and neon—and water vapor, evaporated from the oceans, cooling as it rose through the atmosphere to form clouds and eventually condense into drops of rain.

Solar energy—the energy of the Sun's light—also churned Earth's atmosphere into turbulent motion. Much more energy from the Sun falls on the tropic regions of our planet than on the polar regions, thanks to the tilt of Earth's axis. The hotter air rises; cooler air moves in to take it place. We feel these motions as winds, and it is the Sun-driven winds that carry weather across the face of our world, bringing rain and snow, warmth and cold, as the seasons change.

Rain and wind erode rocks. They wear the rocks down into dust, and much of this material is carried by streams eventually into the oceans. When they first formed, the oceans were made of fresh water, as fresh as any sparkling mountain lake. But as centuries turned to millennia, and millennia to eons, the continuous growth of salts and metals and other elements washed from the rocks turned the oceans saltier and saltier.

The seas and oceans were turning into what biologists now call an "organic soup": that is, a mixture of water and organic compounds, compounds rich in carbon and the other elements that would one day produce life.

All of Earth's living creatures, from viruses and amoebas to whales and sequoia trees, require three basic ingredients: a building-block element that can form complex long-chain molecules, a medium in which these molecules can grow and interact, and a source of energy to drive their growth and interactions.

On Earth, the building-block element is carbon. Carbon atoms can link themselves together into long molecular chains and form extremely complex molecules such as amino acids, proteins, and the elegantly beautiful double helix of deoxyribonucleic acid, DNA, the blueprint of life.

The medium for life is water. No earthly creature can exist without water. We carry a copy of Earth's ancient seawater in our blood vessels. Human blood has about the same salinity as the Earth's oceans did more than 3 billion years ago when life first appeared. For all those myriad generations, life has faithfully reproduced the environment in which it first arose.

The energy for life is light. Without light, life would never have come into existence. Without light, life would be snuffed out completely.

What happened, as nearly as scientists can piece together the story, is this:

In that rich organic soup that made up the seas and oceans of Earth more than 3 billion years ago, long-chain carbon molecules grew more and more complex. Biologists call this *chemical evolution* and have traced in their laboratories the ways that simple molecules tend to combine under favorable conditions to produce larger, more complex molecules.

Critics of the concept of evolution claim that the likelihood of the necessary chemicals combining in precisely the right way to produce life, merely through the blind workings of chance, is so astronomically remote that there must have been some kind of supernatural guidance involved in the creation of life. But as any high school student who has suffered through Chemistry 1 knows, chemicals do not always combine at random; they are usually very specific about how and under what conditions they will combine.

In 1953, at the University of Chicago, Harold C. Urey, a Nobel Prize–winning chemist, and his graduate student Stanley L. Miller showed that chemical evolution was not only possible in those primeval times, but likely. They filled a glass flask with the ingredients of Earth's primitive atmosphere: methane, ammonia, water, and hydrogen. Then they supplied energy in the form of an electrical spark. The gaseous mixture was then circulated through a water bath.

The double helix of DNA, key to all life on Earth. At the heart of every cell of your body is a DNA molecule that contains the information to precisely reproduce that cell.

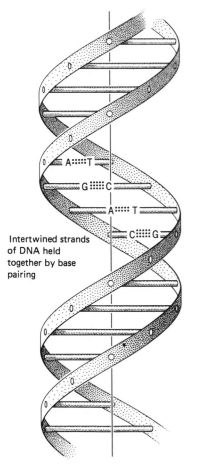

Intertwined strands of DNA held together by base pairing

After a week the water had turned dark brown. Chemical analysis showed that the experiment had produced several amino acids, long-chain carbon-based molecules that are the building blocks of proteins.

What scientists can do in a week with a few quarts of chemicals, nature could certainly have accomplished over millions of years with whole oceans as her laboratory. Miller and Urey used electrical energy as the driving force in their experiment. Perhaps back in those early seas, lightning flashing from the clouds helped to strike the spark of life. Even so, it was the steady dependable energy of sunlight that gave life the ability to grow and flourish.

At this stage of life's earliest beginnings, light was both a friend and enemy. Energy was needed to make longer and more complex molecules. But too much energy, especially in the form of ultraviolet light (the rays that tan), could break up big molecules.

Most likely it happened in shallow seas, close to shore, where sunlight could warm the water but ultraviolet light could not penetrate the water to undo what the gentler energies of sunlight were accomplishing. Or perhaps it took place in clay banks along the water's lapping edge. There are many possibilities, and the evidence was wiped away eons ago.

Long-chain molecules grew more and more complex, adopting spiral forms, adding smaller molecules to their chains, growing and adapting to the kindly environment of the shallow waters. Some of these molecules acquired the ability to reproduce themselves from the simpler organic molecules floating all around them.

Life began.

Life is, chemically speaking, a molecule that can reproduce itself out of simpler building materials. That is the difference between living creatures and nonliving things. A molecule of a living creature can reproduce itself from molecules that are simpler than itself.

A fire eats and grows and eventually dies, but it is not alive. It breaks down the stuff it feeds upon; it turns candles into melted wax and heated air, forests into charred lifeless stumps. Life builds. Life grows. Life can take simple inputs such as sunlight and water and carbon and create an orchid. Only life can do that.

By their fruits you shall know them. The end result of life is more life.

The earliest living things were far from orchids. Or even amoebas. They were microscopic collections of atoms, perhaps as big as a thousandth of an inch in length. They had developed the ability to link with smaller, simpler molecules and rearrange this "food" into copies of themselves.

All life on Earth today is based on the DNA molecule, the double helix that contains the genetic information for replicating each organism. In the nucleus of every living cell on this world is DNA, holding its precious blueprint, unwinding to yield up the chemical information that builds more copies of the parent cell.

Perhaps the very earliest living organisms were based not on DNA but on some other molecular blueprint of replication. All we know for certain is that eventually DNA triumphed over any competitors. *All* of Earth's creatures, plant and animal, from protozoa to primates, from algae to sequoias, are based on DNA.

(Certain viruses contain no DNA, but they can only reproduce by inserting themselves into living cells and inducing the cells' DNA to make copies of the virus.)

The earliest organisms of Earth were all animals.

Not in the modern biologist's sense. But these microscopic molecular life forms could continue to exist only by "eating" the organic molecules around them. Biologists call them *heterotrophs*, creatures that must ingest organic compounds to live. You and I are heterotrophs; so is a paramecium, a mosquito, a Bengal tiger.

The earliest heterotrophs had to "eat" other organic molecules to stay alive. Thus they faced a crisis right from the start, although they could not know it. The crisis was this: What happens when all the "food" in the organic soup is used up?

They were, of course, totally unaware of the impending crisis. To these earliest forms of life, the organic soup in which they lived was a Garden of Eden: Food was there for the taking, plentiful and easy to come by. How quickly these molecules evolved into more complex organisms is unknown. Biologically, the step from a self-replicating molecule to a single-celled creature like an amoeba is a far greater leap in complexity and organization than the step from amoeba to human being.

Most likely the first molecules of life gradually evolved into microscopic spherical shapes, called *proteinoid microspheres*, before the development of true cells. But while this evolution was taking place, the food supply in the organic soup was dwindling.

Undoubtedly these first heterotrophs were gobbling up their

organic food faster than the geologic forces of erosion could stock the seas with fresh materials. If the creatures ate all the available food and could find no more, they would then die off. Of course, they might be able to cannibalize one another, but sooner or later there would be just one cannibal left, and quickly enough it would die too. Life would end, a brief candle that starved to death.

Sunlight. The energy of sunlight saved life from early extinction.

It must have been a time of wild experimentation among the greedy gobbling creatures in those sun-warmed seas. Unknowing, uncaring, they had a whole world of food and energy to play with, and they must have multiplied and diversified enormously.

They left no evidence, those first ancestors of ours, for they were merely microscopic collections of atoms. No hard shells to petrify and be found eventually by geologists. Feathers and bones and teeth, the kinds of clues that paleontologists deal with, would not develop until billions of years later. These tiny creatures lived, died, and dissolved without a trace. Yet they must have been multiplying as fast as they could.

And changing. Constantly changing. In modern biology laboratories scientists see viruses, bacteria, even creatures as complex as fruit flies, undergoing mutations and changing from one short-lived generation to another. The thriving organic soup of those primeval days must have been a biological laboratory par excellence. Mutations came and went. One generation of hungry creatures followed another in minutes.

Somehow, somewhere, at least one of those earliest creatures developed a molecule that could use the energy of sunlight to produce food for itself. Sunlight, water, carbon dioxide, and simple *inorganic* elements were all it needed to sustain itself. It did not have to gobble other living creatures; it did not even need organic (carbon-chain) molecules to sustain itself. It could manufacture its own organic materials out of sunlight, water, carbon dioxide, and inorganic elements.

The first *autotroph* was born, a creature that could manufacture its own food. A plant, rather than an animal. Its solar-energy-using molecule was not exactly like the molecule of chlorophyll that is at the heart of every green plant cell on Earth today, but nevertheless that molecule could perform the miracle we call photosynthesis.

Photosynthesis: putting together with light. Ever since that supremely crucial moment when the first autotrophs began to use sunlight directly as their source of energy, the future of life on Earth was assured. Food chains developed, in which plant life created food and animal life ate the plants—or one another. Glowing at the apex of the food chain there is always the Sun, beaming its unfailing energy, making life possible.

Life survived its earliest crisis, thanks to light. The dead end of cannibalism was averted. Photosynthesis, the ability to live on light, saved life from extinction.

When did this happen? The evidence is hazy, but the Hungarian-American biochemist Bartholomew Nagy and his wife, Lois Anne, found in the 1970s traces of microscopic round structures in rocks laid down in Swaziland, in southern Africa, some 3.4 billion years ago. No more than a mere few thousandths of a millimeter in diameter, these faint shapes may be evidence of single-celled blue algae, although the Nagys caution that such evidence is very uncertain. The oldest verified evidence of life comes from Rhodesia, where rocks dated at approximately 3 billion years bear *stromatolites*, the fossilized remains of filaments and mats of algae.

Single-celled algae may appear to be lowly creatures to us, but compared with the original molecules of life they are as complex as an automobile factory. (See Plate 1 following page 78.) If such algae existed in colonies 3 billion years ago, and perhaps as individual units nearly half a billion years earlier, then life must have originated closer to 4 billion years ago, soon after the crustal rocks cooled enough for pools of water to form.

It looks as if life, tenacious and opportunistic, arose as soon as the ingredients for its creation came into being.

Not only did the energy of light help in the creation of the first living organisms. It was the development of photosynthesis—the ability to sustain life on the energy of light itself—that allowed life to survive and flourish.

More than that, though. As life developed it began to change its world. A subtle, beautiful symbiosis began in which life adapted itself to its surroundings while at the same time it altered its surroundings to make way for more life.

3

As Clear as Air

THE MIRACLE of photosynthesis not only assured the success of life on Earth, it also drastically changed the Earth's atmosphere and prepared the way for life to leave the seas and move onto dry land.

But it took 2.5 billion years for that to happen.

The earliest photosynthetic creatures were not true plants, not as a modern biologist would define the term. They may have been little more than a collection of long-chain molecules or the next step up on the evolutionary ladder, proteinoid microspheres some four ten-thousandths of an inch in diameter.

The evidence shows that by 3 billion years ago true cells had evolved. A cell is a very complex organism, compared with even the largest and most complicated molecule. Cells have specialized parts within them for digestion, for motion, and even for reaction to light. Biologists have speculated that the earliest cells came about when individual organisms joined together, perhaps accidentally at first, and began to take on specialized tasks (such as digestion) within the new group and leave other tasks (such as locomotion) to other members of the group.

Thus the first cells may have represented a sort of symbiosis, a cooperative arrangement of organisms that were once independent and free-living. By banding together and specializing, they could live longer and produce more offspring. "Co-op living" superseded "rugged individualism" at the microscopic level of life more than 3 billion years ago.

The earliest creatures to use photosynthesis existed in the warm waters of Earth's primeval seas. The atmosphere above those seas was still a mixture of nitrogen, carbon dioxide, ammonia, and methane for the most part. Unbreathable to us because there was no free oxygen in it.

The photosynthetic creatures were not bothered by the lack of oxygen. In fact, oxygen was to them a deadly poison. The descendants of these creatures are still among us; biologists call them *anaerobic bacteria*, meaning that they exist in environments where free oxygen is not present. Anaerobic bacteria are the little beasties that live in turgid swamp water or in dead tissue. They can cause tetanus and gas gangrene. Another of their nasty tricks is to get into canned foods and cause botulism.

Let's put it this way: Anaerobic bacteria are not good for oxygen breathers. And vice versa. Oxygen kills them.

These early photosynthetic bacteria thrived at a time when there was no oxygen in Earth's atmosphere that was not chemically combined into compounds such as carbon dioxide. (Or, at best, very little free oxygen.) However, they did not use the sunlight that came through that oxygen-poor atmosphere very efficiently.

Sunlight is composed of a rainbow of colors, which can easily be seen by letting a ray of sunshine pass through a glass prism. The spectrum of visible colors ranges from violet through blue, green, orange, yellow, and red. Beyond our ability to see them lie additional, invisible "colors," such as the tanning rays of ultraviolet and, at the other end of the spectrum, the rays of infrared that our skin senses as heat.

In the cells of the photosynthetic bacteria were pigments that absorbed light and made photosynthesis possible. These pigments were originally nothing more than chemicals that absorbed certain wavelengths of light and stored that energy in chemical form. The pigments of the photosynthetic bacteria absorbed mainly reddish light, which is less energetic than light that is more toward the blue end of the spectrum.

Nature rewards efficiency. Other single-celled creatures developed pigments that could absorb the more energetic, bluer light. We know them as the blue-green algae. Their pigments were built around the chlorophyll molecule, and they could make much more efficient use of sunlight than the photosynthetic bacteria. With

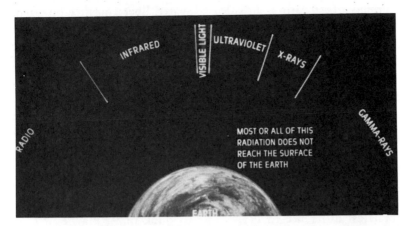

Sunlight consists of a rainbow of colors, which scientists call a spectrum. Beyond the visible colors are wavelengths of light that our eyes cannot detect, such as the ultraviolet rays that tan our skins and the infrared radiation that we sense as heat.

chlorophyll, they could build more of themselves out of sunlight, water, carbon dioxide, and inorganic chemicals. And they could do it more efficiently than the anaerobic photosynthetic bacteria.

But there was a terrible price to be paid for this efficiency. The blue-green algae gave off oxygen as a waste product. Deadly oxygen was let loose into the waters, where it bubbled up to the surface and escaped into the atmosphere.

Oxygen is an extremely corrosive element. It reacts powerfully with other chemicals. It will rust steel and feed fire. All of which is a way of saying that oxygen is an efficient chemical activist.

Nature rewards efficiency.

The blue-green algae had no intention of changing the world. Like all organisms they were merely groping blindly for an ecological niche where they could thrive in safety. Their more-efficient use of sunlight gave them an edge on their bacterial competitors. Their by-product of oxygen had the genocidal side effect of driving the anaerobic bacteria almost to extinction.

And of changing the world.

There had been oxygen in the seas, dissolved in the water. It provided the input for the organisms that were not chlorophyllic, the heterotrophs that evolved into the molluscs and arthropods and fishes that swim the oceans. But the chlorophyllic organisms were producing more oxygen than the seas could hold. For millions upon millions of years their waste-product oxygen bubbled up out of the seas and into the atmosphere. Gaseous oxygen reacted strongly with the methane and ammonia in the atmosphere, turning it into carbon dioxide and water, liberating free nitrogen. The carbon dioxide already in the atmosphere, as well as that produced by oxygen's reaction with methane, provided the feedstock for the chlorophyllic algae and their kin.

Over the slow course of eons, Earth's atmosphere changed. Carbon dioxide almost disappeared entirely; today it represents less than 1 percent of the air we breathe, constantly replenished by the exhalations of the world's oxygen-breathing animals, constantly consumed by the inhalations of the world's oxygen-producing green plants.

Methane and ammonia were completely erased from the atmosphere, except for tenuous traces so rare that they make little practical difference to life. Come to think of it, maybe *little* is too small a word. Methane is a good absorber of certain wavelengths of infrared radiation. It is a major constituent of the gas released by decaying plant matter and when an animal "breaks wind." Methane exhalations from the bacteria in the guts of the world's creatures, particularly from the world's teeming billions upon billions of insects, may be slowly altering the world's climate by absorbing infrared radiation and trapping it in the atmosphere in a sort of disgusting greenhouse effect.

Be that as it may, by some 500 million years ago Earth's atmosphere had been transformed by the oxygen-exhaling green plants of the oceans. Methane, ammonia, and carbon dioxide were practically wiped out. The atmosphere now consisted primarily of nitrogen and oxygen, with traces of noble gases (argon, neon, etc.), carbon dioxide, and water vapor. The carbon dioxide was constantly renewed by the oxygen-breathing animals of the sea, although quite a bit of it was absorbed in the oceans and gradually locked into the Earth's rocks in the form of carbonates.

The lowly algae and their descendants had prepared the world for a new burgeoning of life. Oxygen-breathing animal life could expand from its original home in the seas and move onto the land.

Until 1987 scientists assumed that once the Earth's atmosphere had been transformed into a nitrogen/oxygen mixture, the ratio of nitrogen to oxygen was quickly fixed at about 80 : 20. The air we breathe today contains roughly 78 percent nitrogen, 20 percent oxygen, and 2 percent noble gases, carbon dioxide, and water vapor. Samples of air trapped in Antarctic ice 160,000 years ago give the same ratio of nitrogen to oxygen.

However, geologists Robert A. Berner of Yale University and Gary P. Landis of the U.S. Geological Survey found that air bubbles trapped in amber that formed from tree sap some 80 million years ago contain about 30 percent oxygen. Did the atmosphere contain considerably more oxygen in the distant past than it does today? If so, how did the oxygen in our air dwindle down to 20 percent? If the amber samples are valid and the oxygen level really has changed, did this change cause the demise of the dinosaurs, which happened some 60 million years ago? Is our supply of oxygen still dwindling? If so, why? And how fast?

Be that as it may, once oxygen became a major constituent of the atmosphere, it did one other thing to make the landmasses of Earth hospitable to life. As oxygen began to build up in the atmosphere, the usual two-atom molecule, O_2, began to absorb the energetic ultraviolet wavelengths of light coming in from the Sun to form three-atom molecules, O_3, of ozone. Ozone is poisonous to breathe, but in time a layer of ozone molecules built up high in the atmosphere, some 30 miles above Earth's surface.

This ozone layer effectively blocked the most energetic and biologically damaging ultraviolet from reaching the ground. Before it was built up, any creatures attempting to survive outside the protection of the seas, whose water absorbs ultraviolet effectively, would be exposed to harsh ultraviolet radiation.

Could life have moved onto the landmasses in spite of the ultraviolet radiation reaching the surface? It seems doubtful. The actual paleontological record shows that life moved landward only after the ozone layer had formed and provided a "sunscreen" to protect the land from the damaging ultraviolet.

Today, of course, pollutants pumped into the atmosphere by civilization appear to be leaching away the protective ozone layer. (See Plate 2 following page 78.) As that protection of the ozone weakens, skin cancers and other forms of biological dangers will increase. It seems ironic that the by-products of the most modern technologies are working to undo the by-products of the ancient lowly algae.

While life was using the energy of light to literally transform the Earth's atmosphere, living organisms began to use light for another purpose: vision.

It was a tough world out there in those primeval seas. The Garden of Eden that was the original organic soup quickly became an Armageddon of predators attacking prey. The driving forces that pushed our molecular ancestors to develop into cells, and later drove the cells to evolve into multicellular organisms, were primarily twofold: (1) the need to find food; and (2) the need to avoid becoming some other creature's food.

For the photosynthetic creatures, food could be created wherever they could reach sunlight. But creatures who could not manufacture their own food were constantly on the prowl, ready to ingest a fat little autotroph wherever one could be found. And the heterotrophs, the creatures who lived by eating other creatures, were constantly in danger of being eaten themselves.

How to protect oneself against the dangers of this merciless environment? Our primeval ancestors developed methods of sensing the world around them, techniques for gaining some early warning of the presence of danger—and food. And once sex was invented, they naturally needed to sense the presence of a potential mate.

One sense was touch. But if you wait until a predator is close enough to touch you, probably you are going to end up as that predator's dinner. Much the same situation applies to the sense of taste; if you are close enough to taste it, it is close enough to taste *you*. Then there was smell, the ability to sense molecules drifting past and identify them chemically. That is a longer-range sense, capable of providing information from a source that is some distance away. Sharks can sense blood in the water from miles away, for instance, and literally "follow their noses" to their prey. The sense of smell, or its predecessor ability to sense chemicals in the water, was probably the earliest sense developed by living organisms.

Hearing is a longer-range sense than even smell, and in a watery environment, where sound carries very well, it is an extremely valuable and important one. Modern cinema audiences are thoroughly familiar with submarine films and their sound effects of sonars pinging away dramatically. Modern fishes and pelagic mammals such as dolphins and whales use sound as a major source of information and even as a medium of communication.

But of all the senses developed by all the organisms in all the world, vision is the longest in range and the richest in information content. In their book, *Life: An Introduction to Biology*, G. G. Simpson and W. S. Beck point out that "as a stimulus for giving information about the environment, light is in a class by itself in the amount of information it can give and in giving information about things at a distance from the organism."

The very earliest creatures certainly did not have a sense of vision. They could not form images and see the world around them. All they could do was to react to light the way a scrap of paper reacts to a puff of wind—move either toward the light or away from it.

But that was a beginning that led to much greater things. From that humble "digital" reaction of moving toward or away from light came, in time, Claude Monet's paintings of haystacks and the Mount Palomar telescope.

The very pigments that led to photosynthesis probably also allowed those earliest creatures to react to light. Some single-celled organisms have light receptors, sensitive spots of pigment that absorb light. They give no real information about the environment around the creature, except to indicate the presence or absence of light. They can help guide a microscope swimmer toward the light it needs for photosynthesis, for example. They might also serve to warn the creature when a shadow falls across its path.

This simple ability to sense light or its absence gave these early organisms a kind of digital response capability. Light or dark. Go or no-go. Black or white. Not a very varied repertoire of responses, but it had significant survival advantages. And it was the first step toward vision.

Such light-absorbing pigments can also provide information on the *intensity* of light. Single-celled organisms can react to increases or decreases in the intensity of light reaching them. Now the organism's repertoire of responses becomes a little richer. It can sense shadings of light. It can move toward a brighter source of light and away from one that is not so bright. Or vice versa.

Light is capable of carrying much more information than simply off or on, bright or dull. As life evolved toward more complex forms, with single-celled creatures developing into multicelled ones, specialized groups of cells began to take advantage of the information-carrying nature of light. Organisms began to develop eyes.

Some organisms found other uses for light as well. In addition to receiving light, they created it.

What is lovelier than a warm summer evening in the country-side, with the winking glow of fireflies turning the darkness into a wondrous display of soft moving lights?

Or picture yourself in a small boat in the dark moonless night of a tropical bay. You let your hand trail in the water, and a thousand sparkling tiny lights spring up where your fingers drift through the gentle waves.

Or imagine yourself deep below the ocean's surface in a world where sunlight never penetrates. But there is light in this benthic darkness. Schools of fishes display eerie glowing lights, like squadrons of night-flying planes showing their safety lights. Many-armed squids and other fishes glide past, flashing lights from pouches beneath their eyes or rows of spots along their sides. (See Plate 3 following page 78.)

Luminescence. Giving off light.

Some chemicals are naturally luminescent. Phosphorus, for example, glows faintly because it is slowly oxidizing, burning gradually in the oxygen of our air. That is *chemiluminescence*.

In the world of plants, many species of microscopic bacteria and protozoa, even certain kinds of funguses, are luminescent. Among animals, some species of insects, crustaceans, squids, and fishes have developed luminescent organs. This is called *bioluminescence*. No land-dwelling animals, except insects, exhibit bioluminescence.

One biologist thought he had discovered a luminescent species of frog, but it turned out that the amphibian's belly was glowing faintly because the frog had just finished eating a dinner of fireflies.

Tiny sea-dwelling creatures can secrete luminescent liquids into the water when they are disturbed. In Puerto Rico there is a breathtakingly beautiful Phosphorescent Bay where these microscopic invertebrates twinkle whenever you move your hand through the water. In the Pacific, the wakes of ships glow on dark nights from the bioluminescence of such creatures.

Deeper in the sea, some species of squids and fishes have specialized cells that light up on command of the animal's nervous or hormonal system, like headlamps of an automobile being turned on. One type of benthic anglerfish extends a lure from its mouth that lights up in those abyssal depths, drawing its prey (like moths to a flame) toward the angler, who snaps them up.

While the larvae of some insect species are luminescent, and thus are called glowworms, it is the firefly that intrigues us most. The firefly's light is actually used for signaling, to attract a mate. The males fly around, flashing their lights in a coded signal. The females, sitting in the grass, flash their lights in response. It's a sort of bioluminescent dating service, with the flashing lights serving as the communications link.

Walt Kelly, creator of the cartoon strip *Pogo*, offered an alternative explanation for the firefly's flashing light. He had a male firefly tell Pogo that when the light is turned on, the male locates a female.

"And when it's off?" Pogo asked.

The firefly replied, "We sneak up!"

The firefly's flash—in fact, nearly all bioluminescent light—is "cold light" in two senses. First, the light is yellow-green to blue, cool colors as opposed to the warmer hues of the rainbow. Second, bioluminescent light is created practically with no heat production at all by the oxidation of a pigment called *luciferin* and the catalytic action of an enzyme called *luciferase*. Lucifer, the name of the brightest angel of them all, means "shining."

The light-producing cells in squids and fishes often come equipped with lenses and reflectors to direct the light more usefully. Much the same evolution took place with other cells that receive light: Tiny spots of light-sensitive pigment became, in time, sharp-focusing eyes.

4

The Eyes Have It

VISION IS so much a part of us that most of us take it for granted—except when we bark our shins on the furniture when groping through the house in the dark.

Yet to be a person of vision is a good thing, while someone who pushes ahead blindly does not usually win our approval. In his book, *The Story of Man,* anthropologist Carleton S. Coon counted "sharp-focusing eyes" among the five physical gifts that have allowed human beings to rise to their present dominance of this planet.*

The development of the eye hinged on a vital physical phenomenon: Light energy can trigger a nerve into firing a burst of electrical signals. That is the key to vision.

In the human body, our nervous system is constantly sending information from the environment toward the brain, while the brain sends commands to the various organs and parts of the body. This information is carried by flickering bursts of electricity fired by individual nerve cells, which are called neurons. The total amount of electrical activity going on in your brain amounts to only about two one-hundredths of a watt, which means it would take a thousand people's brains to light a 20-watt bulb. Yet although the brain's electrical activity is low on sheer power, it is very high on sophistication. It is so constant and so complex that

*The other four gifts are erect posture, free-moving arms and hands, large brain, and speech.

it makes even the biggest supercomputer look like a Tinkertoy by comparison.

In those teeming seas of 3 billion years ago, organisms developed the ability to react to light. Many different kinds of organisms have developed true vision, the ability to form visual images of the environment around them. Eyes, "the better to see you with, my dear," have been developed by shellfish and squids, insects, snails, spiders, and almost all the vertebrates, from the jawless fishes of half a billion years ago to *Homo sapiens* in all our multihued splendor. The fact that so many different kinds of creatures have taken advantage of light and developed vision independently of one another shows how valuable and important vision is to survival.

Many species of protozoa, creatures such as amoebas and paramecia, have a special region in their single cells that reacts to light. This photosensitive region contains a pigment that undergoes a chemical change when light strikes it. The pigment is usually a protein that contains a portion of a carotenoid molecule. Carotenoids are pigments sensitive to yellow light that are found in many plants. Animals obtain carotenoid as vitamin A. Since the vitamin can be produced only by carotenoid-rich plants, it behooves us to eat our carrots and other vegetables, especially if we want to have good vision.

From a pigment spot in a single-celled organism, multicellular invertebrates moved a step further on the road to true eyes and vision. Some of their kind developed a rudimentary kind of lens, backed by a collection of light-sensitive cells. The lens concentrates light, so that the cells can react to weaker intensities of light than they would be able to react to without the lens. It also permits a finer discrimination between varying intensities of light.

Such a primitive eye can also yield a completely new kind of information for the organism: It can tell the direction from which the light is coming. If the light falls more intensely on the cells on the right side of the eye, for example, that means the light is coming from the right.

When there are enough light-sensitive cells behind the lens, the eye can even tell if the source of light (or shadow) is moving across its field of view. It can sense motion in that it can discriminate light or dark starting at one side of the field of view and moving toward the other. No image is formed by this kind of primitive eye; the organism cannot tell if the object in motion is a predator or a chunk of food or a potential mate. But it can sense that *something* is moving at some distance from itself.

While lenses originally evolved merely as a means of concentrating more light on the sensitive receptor cells, they also eventually became useful as a means of focusing an image on those cells. Of course, the number of photoreceptor cells had to increase enormously to handle the vastly increased amount of information that imagery entails. The true eye consists of a retina, the curved surface lined with light-sensitive photoreceptor cells, and a lens to focus images on the retina.

The images that the retina receives are turned into nerve pulses and sent on to the brain, where the information is processed and decisions are made about what to do next. In Chapter 5 we will see more detail about this fascinatingly vital process.

One of the earliest steps toward true vision was taken by the lowly planaria, the flatworms. These little fellows are mostly freshwater swimmers, although some live on land and a few species exist in saltwater. They seldom get as large as half an inch in length and have a sort of arrowhead-shaped head, usually marked with a couple of prominent patches of color marking their eyespots. Most of them have two eyespots, although species with four and even six are not uncommon. One species of land-dwelling flatworm, *Geoplana mexicana*, has eyes not only at its head but also strung along the length of its tiny body, like the running lights of a highway semitrailer truck.

It is significant, though, that flatworms (and every other animal with a sense of vision) tend to keep their eyes as close to their brains as possible. Quick as a flash, nerve impulses still take time to go from photoreceptors to brain. The shorter the distance between the two, the faster the organism can react to what it has seen. Milliseconds count. Being quick on the draw was an important survival trait long before Wyatt Earp and Dodge City.

The flatworm's eyespots are mere cups containing light-sensitive pigment. They serve to detect the presence or absence of light, and most planaria move away from light. They are nocturnal beasties and totally carnivorous; their prey are single-celled protozoans,

The lowly flatworm was among the earliest creatures to develop a primitive form of vision. Its eyespots can detect the presence or absence of light but cannot form images, as human eyes can.

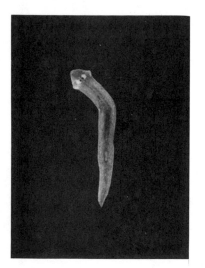

tiny snails, and worms down in the vicious microscopic jungles that we see as lovely gardens and tranquil ponds.

Biologists have used the flatworm's sensitivity to light to study how nervous systems work. They put planaria in watery mazes and observed their reactions to light and other stimuli. For years a group of scientists engaged in this research published a mostly-serious journal called *The Worm Runners Digest*.

Molluscs developed true vision. Some species of snails, certain shellfish such as the scallops, the squid, and the octopus have all developed eyes with real lenses and sophisticated retinas. It is rather startling to see a row of bright blue eyes peeking out from the edges of a scallop's shell. Even the largely immobile barnacle has simple eyes. When a shadow passes over a barnacle it immediately withdraws its feeding cilia and clamps its shell tightly shut. Charles Darwin was fascinated to find that barnacles reacted to the shadow of his hand passing between them and a single candle.

Spiders have eyes that are remarkably like those of the most advanced snails, yet they must have developed them independently, since the two family lines evolved quite apart from one another.

Insects, of course, developed compound eyes. Thousands of individual lenses, packed close together, are individually connected to photoreceptor cells by tubes rather like a miniature version of the tube of a telescope. However, in the insect eye, there is only one lens, at the top, and the light it gathers is guided down the tube to the receptor cells. With thousands of individual images being carried to the brain, insects are very sensitive to motion in their field of view. This comes as no surprise to anyone who has tried to swat a mosquito.

Many researchers have concluded that the mosaic of images formed by insects' compound eyes must be rather crude compared with the finer vision of camera-type eyes such as our own. Yet there is no way of knowing how an insect's brain processes the imagery carried to it. Remember, the visual images received by the photoreceptor cells are converted into bursts of electrical nerve impulses and sent on to the brain as such. What happens inside the brain to transform those pulses of electricity into "pictures" of the surrounding environment is still very much a mystery, whether we are looking at a house fly or a biologist.

The human eye is among the best of the image-forming type that biologists have taken to calling "camera-type" eyes in a sort of reverse wordplay that delights etymologists.

The human eye is shaped like a globe. An adult eye is about one inch in diameter. At birth, a baby's eye will be a bit more than half an inch across; it grows to about 90 percent of its adult size by the time the baby is three and reaches its full size by the age of 13. The adult eye weighs about a quarter of an ounce.

Interestingly, the eyes of the largest animals on Earth are not very much bigger than human eyes. A 90-foot-long blue whale, for example, may have eyes that are about three inches in diameter. That appears to be good enough to gather in all the light the whale needs to see by. Like us, the whales are mammals; they forsook the land to become ocean denizens some 50 million years ago.

Our eyes have lenses that focus light sharply on our retinas. The lens is covered by a tough, transparent cornea to protect it and by an iris that controls the amount of light allowed into the eye by dilating or contracting, exactly like the diaphragm of a camera (which was derived from the eye's diaphragm, the iris). Between the cornea and the iris is a clear liquid called the *aqueous humor*.

The iris is the part of the eye that is colored, and eye colors range from the icy blue of Norsemen to the rich brown of Nigerian Ibo to the breathtaking violet of Elizabeth Taylor. (See Plate 4 following page 78.)

A simplified diagram of the human eye, called a "camera type" eye because it can form images on the retina.

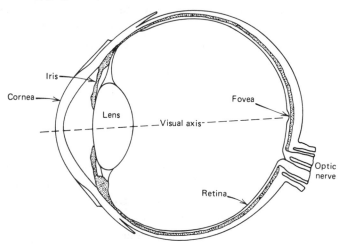

The opening in the center of the iris is called the pupil, and it is through the pupil that light enters the eye, passes through the lens, and falls onto the retina. The pupil appears black, no matter what the color of the iris, because it reflects practically no light at all. The light goes through the pupil and into the eyeball. As the iris dilates or closes, the pupil seems larger or smaller. The iris reacts not only to the amount of light it receives but also to the emotional and chemical environment of the body. Fear or surprise can narrow the iris or dilate it; narcotics also manipulate the iris, sometimes making it so dilated that the drug user needs dark glasses to protect against the painful sensitivity to light.

When a person is in a dark room, the pupil can widen to nearly a third of an inch in diameter (approximately eight millimeters). But turn on the lights and the pupils immediately constrict. Even if only one eye is exposed to light, both pupils constrict at the same time. In bright light the pupils may stay only 0.13 inch (3.5 millimeters) wide.

Light enters through the cornea before it gets to the pupil. The tough, transparent cornea protects the rest of the eye from the dirt and dangers of the outside world. It also provides most of the focusing of light; the lens on the other side of the pupil handles the fine adjustments that make human vision so precise. With age, the cornea tends to harden and roughen, which can cause astigmatism. In some cases the cornea becomes opaque, and in the past this invariably led to blindness. Modern surgical techniques (including lasers that use intense beams of light in place of sharpened steel scalpels) can remove dulled corneas and return sight with the aid of eyeglasses or contact lenses to replace the focusing and protective functions of the cornea.

Corneas can be transplanted, if suitable donors can be found. In 1986 there were more than 28,000 corneal transplants in the United States alone, more than all other transplant operations combined. Surgeons are now beginning to implant artificial corneas, made of soft plastic, developed at the Tulane University Medical School. The first such implant was performed in December 1987.

Immediately behind the iris is the eye's lens. It is necessary to have some sort of focusing mechanism for the eye if its owner expects to see things both near and far. Without a focusing mechanism, the eye will be able to see clearly only those objects that are at a fixed distance away; everything else will be blurrily out of focus.

In human beings, fine focusing is accomplished by actually bending the lens. Strong muscles attached to the lens bend it slightly, contracting to make the lens thicker for seeing things close at hand, relaxing to make the lens thinner for longer-distance vision.

While mammals in general do their fine focusing by bending the lens, other sighted creatures use different focusing techniques. High-flying predatory birds such as hawks and eagles must be able to discriminate their small-game prey from distances of miles and then keep that mouse or rabbit in focus as they swoop down and take it in their talons. Such birds can drastically change the curvature of the cornea, which produces a much greater range of focusing than changes in the curvature of the lens could.

Molluscs such as the octopus actually flatten out the entire eyeball to bring the lens closer to the retina, then relax the controlling muscles to allow the eyeball to resume its normal shape. Fish move the lens forward and back, rather like extending or pulling back the barrel of a telescope.

Human beings possess not only an extremely acute sense of vision but *color* vision as well. Color adds an extra dimension of sensitivity to our vision; it is much easier to discriminate between two objects that are of different colors than two that are of the same hue. Picture a map of the world all in one color, then visualize that same map with the seas and each nation colored. You can see how useful color is. The same usefulness helped our

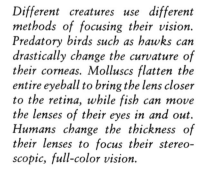

Different creatures use different methods of focusing their vision. Predatory birds such as hawks can drastically change the curvature of their corneas. Molluscs flatten the entire eyeball to bring the lens closer to the retina, while fish can move the lenses of their eyes in and out. Humans change the thickness of their lenses to focus their stereoscopic, full-color vision.

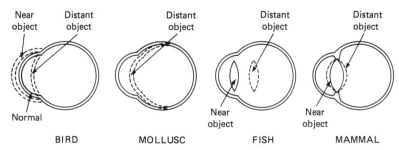

distant ancestors to perceive the world about them with greater clarity and discrimination.

Most mammals cannot see colors very well. A matador could wave a gray cloth at a bull and get the same reaction from the beast. The muleta is red for our excitement, not to stir an angry charge from *el toro*. On the other hand, honey bees can see ultraviolet, a color that is invisible to us, while insects in general cannot see red. To them the color red appears black, just as ultraviolet is "black light" to us.

The gorgeous profusion of colors that the flowers of the world present is not for our appreciation but for the purpose of attracting pollinating insects and birds. In fact, if we could see ultraviolet the way honey bees and several other species of insects can, flowers would look very different to us. Many of them bear ultraviolet stripes that point to their nectar-rich hearts, like the guiding arrows painted on airport runways.

Dogs, cats, horses, cows, virtually all of the animals we have domesticated, have, at best, a very feeble color sense. But our closest primate relatives, the apes and some species of monkey, share our keen color vision. So do many species of birds, insects, and even fishes. Anthropologists have speculated that our distant ancestors, those protoprimates that lived in forests millions of years ago, made use of color vision in finding fruit among the leafy trees and then determining if the fruit was edible. The next time you go to a supermarket and begin to check out the apples or peaches, turning them over to look for bruises or spoilage, remember that you are using your color vision in the same way your distant forebears did.

You will also be using your stereoscopic vision, the ability to focus both your eyes on the same spot in order to determine depth or distance. Stereoscopic vision must have been an important asset for creatures that lived in high trees, where missing a branch as you swung home could mean a screaming, fatal fall.

Those distant days of tree dwelling have left an indelible imprint on our minds. The three most common fears among humans are fear of darkness, fear of falling, and fear of snakes. To a tree-dwelling species, darkness meant danger unless it was safely bundled into a warm, cozy nest. You cannot see where you are going in the dark, and even our early ancestors depended heavily on vision; they were *diurnal* (daylight-active) creatures, not nocturnal animals. Fear of falling is obvious to a tree-dwelling species. Infants display an innate fear of heights at the age of only a few months. And snakes must have been one of the few predators that could reach our monkey ancestors up in their leafy nests. Even in the dark.

Our distant ancestors descended from the trees a few million years ago and began to live on the ground, often on grassy savannas similar to the steppes of central Asia or the prairies of North America. Good vision was crucially important to them as well. They were hunters—and sometimes the hunted, for there were great cats with more speed and strength, bigger fangs and claws, then the puny descendants of tree-dwelling apes. It was important to spot the predators from a long distance away. It was equally important to be able to see meat animals from as far a distance as possible.

In modern humans, when eyesight begins to fail with age it most often results in far-sightedness, hypermetropia. We can no longer focus on objects close to us. We need corrective lenses to be able to read. Normal vision, for humans, is rated as 20/20. That is, you can distinguish the letters on an eye chart held 20 feet away from you; 20/40 vision means that the eye-chart letters

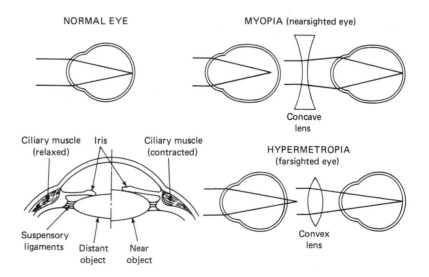

NORMAL EYE

MYOPIA (nearsighted eye)

Concave lens

Ciliary muscle (relaxed)   Iris   Ciliary muscle (contracted)

HYPERMETROPIA (farsighted eye)

Suspensory ligaments   Distant object   Near object

Convex lens

For sharp-focused vision, the eye's lens must focus incoming light on the retina at the rear of the eye. Nearsightedness (myopia) occurs when the eye's lens focuses at a spot short of the retina. A concave eyeglass lens can correct the problem. In farsightedness (hypermetropia) the eye's lens focuses to a spot beyond the retina; convex eyeglasses correct the problem.

look to you as if they were 40 feet away when in fact they are only 20.

An anthropologist might point out that being far-sighted was not as great a handicap to our hunting ancestors as being near-sighted, the condition called myopia. Myopic hunters in the Stone Age stood a much greater risk of not seeing that saber-toothed cat until it was much too close to escape. Even today, we prefer far-sighted leaders to myopic ones!

Four-legged creatures that live close to the ground depend upon the senses of smell and of hearing much more than vision. Their environment is hemmed in by grass and shrubs; it is impossible to see more than a few yards in any direction. Smell and hearing can give earlier signals of danger or food than vision. Dogs and most other "close to the ground" animals have highly developed senses of smell and hearing. A bloodhound can catch a few molecules brushed off the trousers of a fleeing criminal and track him down where no human being could find him. Dogs can also hear frequencies of sound inaudible to humans: hence the usefulness of dog whistles.

Deer, despite their big brown eyes, cannot see colors very well and do not perceive objects that are standing still as threats to them. They may hear and smell a stalking mountain lion, but they will see it only if the lion is moving. They will not notice a man in a bright red jacket with a loaded shotgun if he is standing still.

Humans, though, use vision more than any other sense. Our senses of smell and hearing are picayune, compared with those of dogs, cats, or the animals of the forest. But we can see colors and discriminate objects better than any critter in the woods. This makes us more dependent on vision for long-range information than on any of our other senses.

Soldiers who have fought in the jungle have reported that the inability to *see* an approaching enemy was more nerve-racking than actually being attacked.

Charlton Ogburn wrote in his book about World War II in Burma, *The Marauders:*

The worst thing was the suspense. Of course in the jungle you could never see a thing except a small stretch of the trail ahead.

Humans depend on vision more than their other senses. Soldiers fighting in dense jungles say that the inability to see an approaching enemy is more nerve-racking than actually being attacked.

Not just when it happened but a hundred times a day you lived in anticipation through the sudden *pup-pup-pup* . . . *pup-pup-pup-pup* of a machine gun opening up on the column. . . . Ahead the view was always closed by a bend in the trail. Always there was a bend to be rounded. Each one had to be sweated out . . . *what was around the next bend?*

The brains of animals tell much about which senses they depend upon. In a dog's brain, for example, considerable areas are devoted to the sense of smell. In humans, the visual cortex is larger and more complex than any sense-related area—except for the sense of touch, which is spread over a huge part of the brain. Perhaps that is why most humans close their eyes when kissing or making love: to capitalize on the sense of touch by shutting out the competing signals that vision would bring.

Along the back of the human eyeball lies the retina, which corresponds to the film in a camera. The retina is a thin layer of photoreceptor cells covering the inner surface of the rear of the eyeball, where the incoming light is focused and images are formed. Between the lens at the front of the eye and the retina at its rear is a transparent jelly called the *vitreous body*.

There are two types of cells in the retina, rod-shaped cells and cone-shaped ones.

The rods are more abundant around the edges of the retina; they can detect light that is too feeble to stimulate the cones. The pigment in the rods, rhodopsin, is so sensitive to light that a molecule of rhodopsin can undergo the chemical change needed to send a nerve impulse toward the brain on the stimulus of merely a single photon (the fundamental particle of light). Therefore, one photon striking a rod cell is enough to "see." Such impulses from the rod cells can tell the brain that light is present, but they convey no information concerning color.

The rods are all-important for night vision. Experiments in which subjects were kept in the dark for half an hour showed that dark-adapted eyes are about 10,000 times more sensitive to light than they are normally; the rods take over the task of seeing when incoming light is too dim for the cones to use. This is why everything appears colorless in dim light. The color-discriminating cone cells are out of action.

The cone cells are more plentiful in the central region of the retina. They are capable of sensing well-defined images, and they can detect color. There are at least three functional types of cones, each containing a different pigment, each sensitive to different colors. They are blue-absorbing, green-absorbing, and red-absorbing. Apparently the three types are enough to allow us to perceive all the subtle shadings of color that we see around us.

Other species, however, have four color-absorbing pigments in their retinas. Chickens, pigeons, and the exasperating cockroach all have four pigments. Does this mean that they can perceive color more finely than we can? That is impossible to tell, since *perception* takes place in the brain, not the eye. Yet the clever cockroach, persistent pest that it is, cannot see the color red at all. To a roach, red is the absence of light, blackness. A biologist friend of mine rid his Manhattan apartment of roaches by putting red bulbs in all his lights and then patiently observing the roaches until he found their nest. Then he wiped them out, and they have never returned.

But some people are color-blind: They cannot distinguish certain colors. They may be blind to red, green, or blue; to two of the primary colors or even to all three. Blindness to red is called protanopia; to green, dueteranopia; to blue, tritanopia (from Greek, meaning "first, second, and third lack of vision"). Apparently color blindness stems from a lack of color-sensitive cone cells in the retina or an inability of the cones to transmit their signals to the brain.

Red-blind persons cannot distinguish between red and green; blue-blind cannot distinguish between blue and yellow; green-blind simply cannot see the color green, it appears grayish to them. In modern society, color blindness is more of a nuisance than a danger. Generally color-blind people can go through life quite normally, and those around them will hardly guess that they cannot see certain colors. A color-blind friend of mine has his wife help him buy his clothes, then lays them out in his bureau drawers by colors, so he can dress in as fine a sartorial display as those of normal color vision.

Color blindness affects about 20 times as many males as females. It is a sex-linked recessive trait; a woman must inherit it from both her parents to be color-blind. Daughters of a color-blind man and a woman of normal vision will have normal color vision but will be carriers of the gene for color blindness; they could pass it on to their offspring. Sons of such a couple will have normal vision and will not carry the gene; therefore they will not transmit color blindness to their children.

But there is much more to seeing than merely detecting light. The eyes are only part of the human vision system. The light they receive is transformed into nerve impulses and sent on to the brain. It is within that complex three-pound congregation of little gray and white cells that the stimuli detected by the eyes become visions of the world around us.

5

"The Vision and the Faculty Divine"

A FLATWORM can detect light and react to it. A human being can see the world in beautiful detail and color and can even create imaginary scenes within his or her own mind. This is because the human vision system involves not merely the eyes but the brain as well. The eye collects light; the brain perceives vision.

To understand human vision we must examine not only the eye but also the brain. In fact, neurophysiologists point out that the light-sensitive retina is actually an outgrowth of the frontal lobe of the forebrain.

The rod and cone cells of the retina possess pigments that react chemically to light, releasing chemicals that trigger the nerves next to them to send out pulses of electrical energy along their pathways and into the brain.

The retinal rods and cones are covered by a thin layer of cells that are not sensitive to light—a sort of neural screen made of translucent nerve cells that transmit signals to the optic nerve tract. Light must pass through this screen to register on the retina. However, near the center of the eyeball there is a tiny depression, a pit where this layer of nerve cells is flattened out and presents less of a barrier to the light reaching the retinal receptors. This is the *fovea*, a region in the retina that is densely packed with cone cells. The fovea is seldom more than one millimeter in diameter.

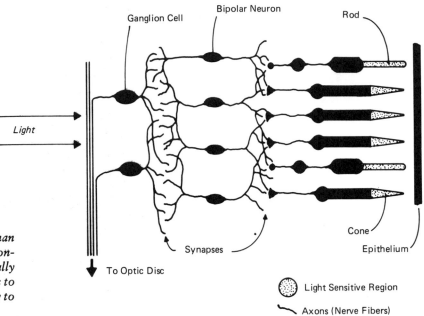

Ganglion Cell   Bipolar Neuron   Rod

Light

To Optic Disc

Synapses   Cone

Epithelium

Light Sensitive Region

Axons (Nerve Fibers)

Cross-section diagram of the human retina. The rod and cone cells contain pigments that react chemically to light, triggering the nerve cells to send impulses of electrical energy to the brain.

(A millimeter is about twice the width of the lead in a mechanical pencil.)

The minuscule fovea is the part of the retina that we use for our most acute vision, our straight-ahead, central vision. When we stare directly at something or someone, we are lining up our foveae on the object of our inspection just the way a battleship lines up its guns on a target. Light reaches the fovea more easily than the other parts of the retina, and the fovea is jammed with cones that distinguish images, as well as color. The rods (apparently more primitive) can detect very low levels of light and are important for our peripheral and night vision. There are no rods in the fovea—only cones, although they are thinner and more like rods than the cones elsewhere in the retina.

The fovea is the central aiming point for our vision, the most sensitive part of the eye. Without deliberate thought, without even knowing that we are doing it, when we look at an object we move our eyeballs so that the foveae aim at it dead center. If the object is too big for the fovea to take it in altogether, we jiggle our eyes back and forth, giving it a foveal once-over. When we read, we move our eyeballs so that the foveae pass over each individual word and letter.

Astronomers have learned that they can see very dim stars if they do not look directly at them. This technique of averted vision works because the fovea contains only cones, which require a higher level of light than the rods, even though they distinguish color and images. The rods, exquisitely sensitive to even the lowest levels of light, are distributed around the fovea. To use them best, the eye must not aim directly at the object to be viewed. Thus to make out a dim pinpoint of light against the dark background of the night sky, the rods work best, and they are not in the fovea at all but spread around the central region of the retina.

There is a blind spot in the retina, the place on the nasal side of the eyeball where the optic nerve leads out of the eye and into the brain. Without this connection the eye could not pass information on to the brain. Yet because of this connection, there is a spot in each retina that contains neither rods nor cones, and therefore cannot perceive light at all. Each of us has our blind spot; without it, we could not see!

There are some 7 million cones in the average human eye and 150 million rods. But only about 1 million optic nerve fibers. (*Only* is a comparative term here. The auditory nerve contains a mere 30,000 nerve fibers.)

This means that each optic nerve fiber must carry a very complex set of signals, rather like a cable in a telephone exchange carrying many conversations at once. This convergence of light signals onto the optic nerve fibers means that each nerve receives an accumulation of stimuli from many rods and cones. Thus a very weak light signal spread out over thousands of rods can be just as effective as a stronger signal received by only a few cells. In effect, the optic nerves add up signals from the plethora of retinal cells, making our vision more sensitive.

Moreover, the network of retinal cells and their interchange with the optic nerve fibers is such that the system tends to respond more to movement and changes in the visual field, such as contours, than to an unmoving, changeless scene. Thus the brain is not bombarded with needless signals. What is important is *change:* Something out there moved, or there seems to be an edge to that brown slope, beyond which is an area of sky blue.

The rod cells in the retina contain the pigment rhodopsin. The three types of cones each contain their own kind of pigment: blue-sensitive cyanolabe, green-sensitive chlorolabe, and red-sensitive erythrolabe.

When rhodopsin or one of the other pigments absorbs light, it undergoes a chemical change. Rhodopsin, for example, is a combination of the protein *opsin* with a chromatophore (pigment-bearing) molecule called *retinal* for obvious reasons. It is the retinal that gives rhodopsin its color. Rhodopsin is also known as visual purple.

Retinal is formed by vitamin A. People whose diets are deficient in vitamin A suffer from degraded vision. When the eye is exposed to light, the rhodopsin in its cells breaks down into retinal and opsin. This photochemical change causes an electrical signal in the optic nerve cells, and quick as a lightning flash the brain perceives the presence of light. The cells regenerate rhodopsin from retinal and opsin, using the body's heat as the energy to drive the reaction.

This regeneration works much better in darkness, where there is no photochemical breakdown of the rhodopsin to compete with the regenerative buildup. Even without darkness, though, rhodopsin is constantly being rebuilt in the cells of the retina, while simultaneously light energy is splitting apart other molecules of rhodopsin. A dynamic balance is established to keep enough rhodopsin available to allow vision to continue. Yet when your eyes get tired you tend to close them for a few moments. This instinctual behavior allows more rhodopsin to be generated. Perhaps sleep is, among other things, the body's way of regenerating rhodopsin for the coming day.

Incidentally, the images that fall on the retina are inverted. Just like looking through the lens of a camera, if we could see the images on our retinas they would all be upside down. That is the physical nature of our optical system, our eyes. These upside-down images are converted inside the brain to "normal," right-side-up images. Where in the brain this happens is unknown, but it obviously does happen. The signals sent into the brain from the retinas are somehow corrected to produce accurate images of the world around us.

How do these pulses of electrical energy become vision? How does the brain take the messages carried by the optic nerves and turn them into discernible pictures?

To understand this, we must first take a look at the brain itself. It is a remarkable organ, a beautifully complex collection of gray and white cells about the size of a grapefruit. The human brain contains about a trillion (10^{12}, or 1,000,000,000,000) nerve cells, which physiologists call neurons. Most of them are only a few millionths of a millimeter in diameter. But they are so intricately interconnected, with each neuron sending out long tentacles to reach as many other neurons as possible, that a microscopic picture of the brain resembles a wildly tangled forest, an impenetrable jungle of neuronal arms (called axons) branching everywhere to connect with thousands of other neurons.

The brain's 1 trillion neurons are interconnected so thoroughly that there must be on the order of 1,000 trillion (10^{15}) links among them. Millions of times more intricate than the most sophisticated electronic computers. Compared with the human brain, the largest and most complex telephone exchanges in the world are mere child's toys.

The brain regulates every breath we take, every beat of our heart, every move we make. Virtually nothing that we do is done without some part of the brain deliberately coordinating it. Almost nothing happens in the body "automatically"; the brain runs the show, second by second, all the years of our lives. While most of our bodily functions are directed without our conscious awareness, there is not a twitch of a facial muscle nor a gurgle of the intestine that is not directed by some part of the brain or spinal cord.

Reflexes such as the blinking of our eyes or the shedding of tears go on without our even thinking about them. And they are beyond our conscious control. They have evolved as mechanisms for cleansing and protecting the outer layer of the eye, the cornea that is exposed to that dirty, often hostile world out there. If we had to make a conscious decision every time we needed to close our eyelids and allow tear secretion to wash off the cornea, we would hardly have time to enjoy what we are seeing. On the other hand, try to consciously prevent yourself from blinking. You might go for two or three minutes, but no further. Your eyes will blink, as they must, despite your most fervent display of willpower.

The structure of the human brain shows the history of its evolution from its humble beginnings as a bunched ganglion of nerves at one end of a primitive sea-dwelling creature. We saw that

Neurons of the brain. The long branching arms are called axons; the electrochemical receivers at the ends of the axons are called dendrites; the tiny gap between axons of different cells is called a synapse. The human brain contains about 1 trillion neurons, interconnected so intimately that there are some 1,000 trillion links among them.

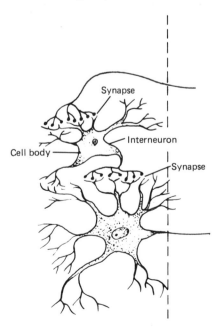

the flatworm's eyspots lie close to its minuscule brain. Did the brain evolve out of the sensory organs that lay at one end of such creatures, or did the sensory organs (eyes, ears, taste buds, odor detectors) assemble near the primitive brain to get their messages through as quickly as possible?

Whichever way it actually happened, the human brain is built upon structures that appeared earlier, in "lower" animals. And it resides in the skull at the top end of the spinal column, very close to our eyes, ears, noses, and tongues.

At the top end of the spinal column, the part that projects into the base of the skull, is the brain stem. It is apparently the oldest part of the brain, and is often referred to as the reptilian brain, since it looks rather like the brain of a reptile. Very basic functions such as heart rate and breathing are controlled by the brain stem. In the center of the brain stem is a core of neurons called the reticular activating system, or RAS.

Sensory information, such as the visual signals coming in from the retina, goes through the RAS, which apparently serves as an early warning system to the "higher" centers elsewhere in the brain. Movements of the eyeballs are controlled partly in the brain stem and partly in the higher centers of the cerebral cortex.

Just behind the brain stem is the cerebellum, which is responsible for motor activities such as balance and movement. The cerebellum appears to be the depository for some kinds of memory, particularly memory of learned responses.

A cross-sectional view of the human brain. The visual cortex is at the rear, in what is called the occipital lobe.

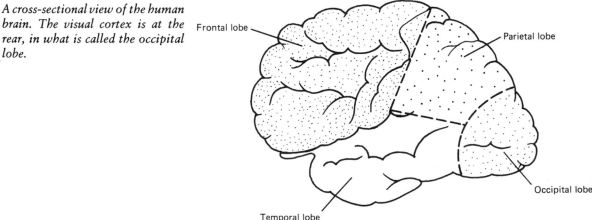

Frontal lobe

Parietal lobe

Occipital lobe

Temporal lobe

Sitting atop the brain stem, deep inside the core of the brain, is the limbic system, a collection of structures that is often referred to as the mammalian brain since it is highly developed in mammals. Here in the area called the hypothalamus, the body's homeostatic systems are controlled. These systems keep our bodies at a constant internal temperature no matter what the temperature outside our skin and direct the release of hormones. The limbic system is intimately involved in our emotional responses, particularly those that are directly connected to survival, such as eating, fighting, and sex.

Right in the middle of the limbic system is the thalamus, a portion of the brain that helps sort out incoming information and route it to the proper areas of the "higher" portions of the brain.

That "higher" part of the brain is the cerebrum and its outermost layer, the cerebral cortex.

By far the largest part of the brain, the cerebrum is a mass of whitish fibers overlaid by an intricately folded layer of gray cells no more than an eighth of an inch thick. This cortical layer, the cerebral cortex, is the most human part of the brain. It is the part that consciously thinks, makes decisions, retrieves memories, reflects upon the past, and visualizes the future.

Agatha Christie's detective Hercule Poirot called the cogitating part of the brain "the little gray cells." While the cerebral cortex's little gray cells are merely a part of the wonderfully complex human brain, they are the most important part to us. The cranial difference between human and chimpanzee is mostly in the cerebral part of the brain.

There are more neurons in the cerebral cortex than in any other part of the brain. Although only about an eighth of an inch thick, the cortex is actually quite large; if one were removed from an average-sized human head and flattened out (a messy business) it would cover roughly six square feet of floor space. Safely inside the skull, the cortex is intricately folded so that it fits neatly atop the twin hemispheres of the cerebrum.

The cerebrum is divided into two hemispheres, left and right, and connected by a thick band of neural tissue called the corpus callosum. For some reason, the left hemisphere of the brain controls the right side of the body, and vice versa. The corpus callosum serves as the connecting bridge that channels nerve impulses from one hemisphere to the other. Sever the corpus callosum of an adult and you get the weird phenomenon of the split brain, where the subject often seems to be in the position of having his right hand literally not know what his left hand is doing.

In the normal brain there are intimate connections between the left and right hemispheres of the cerebrum. Still, a goodly amount of specialization takes place within the brain. The left hemisphere generally is more concerned with language and logic; the right usually specializes in spatial relationships and the kind of intuitive thinking that psychologists call gestalt. But this hemispherical specialization is far from absolute. It is *not* the case that the right side of the brain is all arts and flowers while the left is all symbols and logic. There is constant interchange between the two, and both sides of the brain are active all the time. If either side of the brain is damaged, the other side can take over the lost functions in time—although it may not be able to handle those functions as well as the original hemisphere could.

At the rear of the cerebrum is an area called the occipital lobe. (*Occipital* is derived from Latin and means loosely "toward the back." Most medical and scientific terms are practical rather than poetic.) This is the area where visual signals from the retina are processed; this is the part of the brain that turns those signals into the pictures we see. It is apparently the part of the brain that produces mental images, pictures we conjure within our own minds, visions we see when we dream.

Thus this part of the cerebral cortex, way off in the back end of the brain, is often called the *visual cortex.*

What happens to the light signals that the retina receives as they go completely across the brain to the visual cortex?

To begin with, the light received by the retinal cells is transformed into electrical nerve pulses, as we have seen. Nerves are very sophisticated little electrochemical factories. A neuron will transmit an electrical impulse along the length of its stretched-out axon. When it reaches the end of the axon the impulse triggers the release of certain chemicals that flow across the gap (called the *synapse*) between the axon and the next neuron's receiver (called a *dendrite*). When these chemicals reach a critical intensity, the next neuron fires an electrical signal down its length, and the process is repeated at the next synapse.

The nerve impulses from the eye travel this bucket-brigade way at a speed of barely seven miles per hour; not much faster than an ox cart travels. Still, it takes only a tenth of a second for the information to get from the retina to the visual cortex. Nowhere near as fast as the speed of light, nor even the speed with which an electronic computer operates, but plenty fast enough for our human senses.

The nerve impulses from our two retinas are routed through a connector system just behind the eyeballs called the *optic chiasm*. What the optic chiasm does is to intertwine the signals from the right eye and the left, so that each hemisphere of the brain gets information from both eyes. This intermixed signal goes all the way back to the visual cortex, which is built of stacks of cells in such a way that each portion of the cortex gets signals from both eyes. No part of the brain, from the optic chiasm onward, plays the role of Popeye the Sailor Man. Information from both eyes is shared throughout the system.

Incidentally, the optic chiasm was discovered by no less a man than Sir Isaac Newton, one of the greatest scientists of all time. As

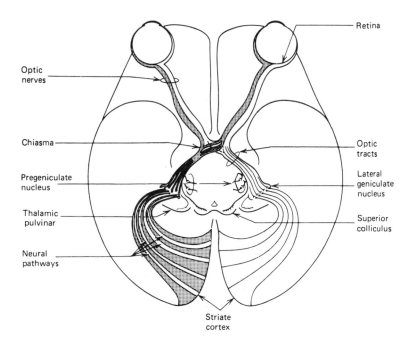

The human visual system. Images formed on the retinas are transmitted as nerve pulses to the visual cortex. The optic chiasm intertwines the signals coming from each eye so that both hemispheres of the brain receive signals from both eyes.

we shall see later, Newton played a vital role in our understanding of optics. While studying the problems of color and vision in 1666, he dissected the visual system of an animal (probably a sheep) and discovered this switching system in the brain behind the animal's eyes.

The optic nerve path goes from the optic chiasm to a structure in the brain called the *lateral geniculate nucleus* in the thalamus, as well as to the RAS and to a part of the midbrain called the *superior colliculus.* The superior colliculus receives signals from the ears as well, and it organizes information about the direction of a moving object so as to coordinate head and eye movements. When you hear a loud noise, for example, you automatically turn your head and eyes toward its source. That is a reflex, beyond conscious control.

From the lateral geniculate nucleus the signals go to the visual cortex, where there are six layers of cells with specialized functions. Some sense colors. Some sense edges, discontinuities in the field of view. Some are so specialized that they detect two edges or lines that connect to form an angle, and little else. It's as if the visual cortex consisted of millions upon millions of highly trained, exquisitely specialized observers, each of whom is responsible for distinguishing one thing and one thing only, no matter what the overall scene being viewed. Put all these special "minipictures" together and we get a splendidly detailed, beautifully colored picture.

Recent experiments performed at the Massachusetts Institute of Technology suggest that even our perception of color depends on *comparing* the light from the object being scrutinized with the light from the objects around it. Not only do we sense colors directly, but we also apparently judge fine gradations in color by comparing all the objects in our field of view against one another. Contrasting shades are much easier to discriminate than shades that blend into one another. This kind of comparison undoubtedly takes place in the visual cortex. The cones of the retina detect colors; the visual cortex compares them and discriminates subtle shadings.

In the visual cortex the incoming information is processed and refined so that all the data about the object being viewed—its size, distance, shape, color, location, and relationship to background— are assembled into a coherent image. Then this information is sent to the *hippocampus* (Greek for "seahorse," which this tiny structure resembles slightly in shape), near the thalamus. The hippocampus is a memory-storage area. Apparently this is for the purpose of checking the new information with old data already in storage: Does this new image resemble anything I've already seen? What category does it fall into? Is it something I should run away from? Or eat? Or smile at and talk to?

From there, still within fractions of a second, the information is spread around the cerebral cortex generally. Take a look at this, the rest of the gray cells are told. What do you make of it? What decisions are necessary to deal with it?

Finally the information is decided upon and the cortex activates the motor controls. The result may be a smile or a frown, a bow or an abrupt dash for safety, a word of greeting or a scream of fright.

Most of what we know about the way the human brain processes visual information comes from experiments on animals such as rats and monkeys, although a good deal of knowledge has been gained through work on people whose brains have been damaged in one way or another. By observing how physical damage to the brain affects vision, it has been possible to pinpoint the areas of the brain that are involved with different aspects of seeing.

There is another facet of human vision that has led not only to scientific knowledge of the way the visual system works but also to a global activity that is part art form and part money-grubbing industry. That aspect is called "the persistence of vision," the fact that if we look at a series of individual pictures flashing at a rate of one each twentieth of a second or so, we tend to "connect" the individual images into a continuous moving picture. This gave rise, around the beginning of the twentieth century, to the motion picture industry and some fifty years later to the television industry.

Flashing lights attract more attention than a steady light. The brain seems to be on constant alert for *change,* which often signifies either danger or opportunity. A person can sleep in spite of the clanking drone of an aged air conditioner, but the instant that noise stops, the person snaps awake. A flashing light also appears brighter than a light of the same brightness shining continuously. The brain is extremely sensitive to changes in the world around us.

When the retina is stimulated by a very brief flash of light, the nerve activity does not end when the light turns off; the neurons remain stimulated for about fifteen- to eighteen-hundredths of a second, no matter how brief the flash of light may be. Almost everyone has experienced the after-image of a flashbulb or an inadvertent glance at the Sun; the after-image glows even when you close your eyes. That is the persistence of vision.

To create the illusion of motion out of a series of still pictures, each individual image must be on the retina for a certain length of time, and the time between images must be brief enough—yet long enough—for the mind to link each image with the one preceding it. The point at which separate images seem to fuse into continuous motion varies with the intensity and color of the light source. Until that fusion point is reached the brain will see each image as a separate picture.

For example, if you were watching a dancer under a strobo-scopic light that was flashing at less than five strobes per second, you would perceive a series of frozen images, "still" pictures in which the dancer does not appear to be moving. Each flash of the strobe would reveal another snapshot of the dancer. At the rate of five flashes per second, there is too much dark time between the strobes for the brain to link up the snapshots. Hence you perceive a series of still pictures, even though the dancer may be gyrating wildly all the time.

Speed up the strobe light's flashes and the dancer seems to be flickering in and out of darkness. But when the strobe gets to about 40 flashes per second, the eye can no longer discern the darkness between flashes; it appears to be continuous light and the dancer seems quite normal.

Motion pictures flash a series of still images on the screen fast enough for the brain to link up each individual image into the appearance of continuous motion. The hand-cranked cameras of the early silent movies resulted in films that were shown at 16 frames per second. At that rate an annoying flicker was present. With the introduction of sound in the late 1920s, the motion picture industry went to 24 frames per second—but films are actually shown at 48 frames per second, because each frame is projected twice. An interrupter device, called an intermittent, blacks out each frame briefly while it is held in front of the projector's lamp. Although the industry standard is stated to be

24 frames per second, when you view a modern movie you are actually seeing 48 frames per second. At that rate, the visual system is quite happy. No flicker, only smoothly continuous motion.

Similarly, although television broadcasts are said to be transmitted so that the entire screen is swept by the picture tube's electron gun 30 times per second, the electron beam actually sweeps the tube 60 times per second. The beam sweeps out every other line on the tube in one pass, then fills in the in-between lines in a second pass, so that the eye perceives 60 sweeps per second and there is no flicker.

Great art has been created in motion pictures and television. Creative artists such as Charlie Chaplin, Orson Welles, Sergei Eisenstein, Federico Fellini, and Akira Kurosawa have enriched our lives. Great nonsense has been perpetrated as well. In a global village in which color pictures and sound can be transmitted around the world with the speed of light, too often we receive images of the village idiot, too rarely the village sage.

The images that the retina receives do not actually overlap, even when they are flashing past at 48 frames per second. Apparently the visual cortex matches one incoming picture with the next, and if the two images are very similar it "connects" them to form a continuous moving image for our conscious delectation. Walt Disney and other motion picture cartoonists offer a clear example of the way this trick of the brain works. Cartoonists draw thousands of individual pictures, each one only slightly changed from the one that preceded it. Flick them at 48 frames per second, however, and we see Roadrunner sprinting away from Wily J. Coyote.

But not all our visions come from observing the world around us. Human beings can create visions within their own minds. We can build imaginary landscapes and play out whole costumed dramas entirely within our heads. Apparently our visual cortex can be stimulated by memories or other stimuli inside the brain to produce scenes that our eyes have never perceived.

Interestingly, though, when we dream—when we build imaginary scenes during sleep—*our eyes move.* Starting shortly after World War II, neurobiologists began using electrodes pasted to the scalp and forehead to detect the electrical activities of the brain's billions of neurons. Those investigating the still-unexplained phe-

Movie cartoons consist of thousands of individual drawings that are projected fast enough for the brain to "connect" them into a smoothly continuous moving picture.

nomenon of sleep found that when the brain's electrical activity indicated that dreaming was taking place, they could see the sleeping subject's eyes moving rapidly back and forth beneath the closed eyelids. REM (rapid eye movement) sleep is now firmly associated with dreaming. We move our eyes when we dream, even though the brain is not using them to peer out at the world.

And we "see" more than the world around us. We can create mental pictures. We can remember scenes from the past or create visions that have never existed. We have dreams in which we see fantastic landscapes while we sleep. Human beings can create mental images, pictures in the mind. "Close your eyes and picture this," says the film writer to the motion picture producer. And we can.

Researchers have suggested that this process of mental imagery may have begun as an aid to seeing the landscape around us, a sort of mental rehearsal for what we might actually see in the real world. A Stone Age hunter might think to himself about the deer he wants to find. Or about the sabertooth cat he wants to avoid. Was this the original motivation behind cave paintings: a subconscious attempt to sharpen the perception of the real world? Is this the emotional drive that produces art: a deeply-rooted desire to shape the world the way we want it to be?

From picking ripe fruit off a tree to watching *Don Giovanni* on television, from cave paintings to the Sistine Chapel ceiling, vision is indeed "the faculty divine," as the poet Wordsworth put it. Our sharp-focusing eyes have helped us to survive the rigors of our beginnings in the Ice Age and have led to an appreciation of the incredible beauties of the world. More than that, human

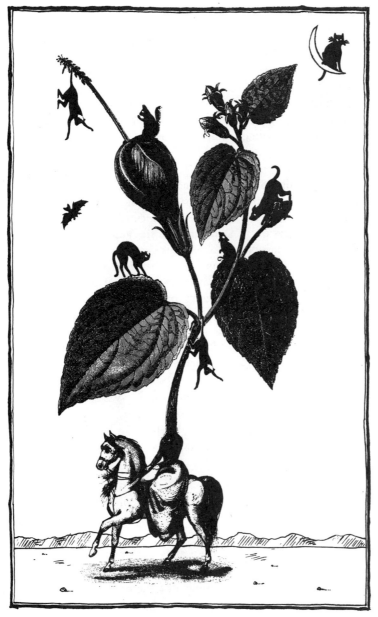

When we dream, our brains create mental images—and our eyes move beneath our closed eyelids. Depicted here is "Dream," by Jean Claude Suarès.

beings have gone on to *create* beauty. Neanderthal hunters buried their dead with flowers. Cro-Magnon hunters painted hauntingly graceful images of game animals on the walls of their caves. Every tribe of humans has created works of art, objects of beauty to be admired.

Rene Dubos, the French microbiologist and environmentalist, pointed out in his book *The Wooing of Earth* that our ancestors' dependence on vision has greatly affected our own ideas of the beauty of nature.

> The experience of several million years of life in an open environment where good visibility was essential for survival and for hunting left a lasting stamp on human nature. During the Ice Age, for example, Neanderthal and Cro-Magnon people settled in valleys rich in game and fish. They took shelter in caves or in dwellings built from branches and animal hides, which were located in places from which animal life could be readily observed. Hundreds of such Old Stone Age settlements have been recognized along the valleys of the Dordogne and the Vezere in France. From the huge Cro-Magnon cave at Les Eyzies, the eye can survey a vast panorama of earth, river, and sky. This diversified environment probably created in *Homo sapiens* two different but complementary types of visual conditioning: on the one hand, the need for open vistas leading the eye to the horizon; on the other hand, the need for a place to take refuge, for example, a cave or a densely wooded area, giving protection in case of danger. Children may be displaying this early conditioning when they play hide and seek.

Neanderthal and Cro-Magnon people were both intelligent members of our own species, *Homo sapiens.* During the time of endless winters we call the Ice Age, these ancestors of ours faced the bitterly cold and dangerous world around them armed with a tool that no other creature on Earth possessed: fire.

After billions of years of using light as the basis of life itself, and adapting to light to develop the sense of vision, the members of one peculiar species of ground-dwelling ape learned how to *make* light in the form of fire.

That discovery changed those apes into masters of the world. That discovery has helped those creatures to succeed so well that their numbers, and the by-products of their fires, pose a threat to the continued existence of life on Earth.

6

The Gifts of Prometheus— and Eve

THE DISCOVERY of fire was so monumental a happening, so incredibly important to the lives of humankind, that every tribe on Earth has constructed a myth to commemorate it.

Not merely *a* myth. The same myth. Go where you will, pick any human culture anywhere on the globe. Dig into its ancient myths far enough and you will find the fire myth. It is always the same story:

Man was a weak, cold, hungry miserable creature, little better than the animals of the field, struggling to survive against the more powerful beasts and the cold of winter. A god took pity on man, stole fire from the heavens, and gave it to those early human beings. The other gods became angry and punished the gift giver, while man—with fire—went on to become master of the world.

In the Germanic and Norse fire myth, the gift giver is Loki. To the Plains Indians of North America, it is the tribal totem, coyote or rabbit or eagle. To the ancient Greeks, progenitors of our Western civilization, the fire bringer was Prometheus, and we tend to call all these myths about the discovery of fire by his name: Promethean.

In Pierre Grimal's *The Dictionary of Classical Mythology*, the

The myth: In Greek mythology, it was the godlike Prometheus who gave fire to humankind.

Prometheus myth begins with Prometheus playing a trick on Zeus, chief god of Olympus. The two were cousins, Prometheus being the son of Japetus, a Titan, while Zeus was the son of another Titan, Cronus.

While some versions of ancient Greek mythology say that Prometheus created man out of potter's clay, most versions of the myth portray him as a benefactor of man, not our creator. The trick he played on Zeus was this: Zeus was to decide how man should properly sacrifice to him. The sacrifice was to consist of burnt offerings—some part of a meat animal would be burned on an altar and its smoke sent toward the sky. But which part of the animal should be burned?

Prometheus cut up a bull into two parts. One part contained the flesh and innards wrapped in the skin, and atop this Prometheus placed the animal's stomach. The other part was nothing but the bones covered with some fat. Zeus chose the fat, and when he discovered that he had picked nothing but bones, while humankind got the real meat, he became so angry that he decided to withhold fire from mortals.

Without fire, humans would freeze and die. So Prometheus stole some sparks of fire from the Sun and gave them to humankind. The gift of fire was to cost Prometheus great pain, however, because Zeus became furiously angry at his cousin.

With fire, Zeus raged (and the other gods apparently agreed), mortals would become as powerful as the gods themselves. There was no way that Zeus could take the gift back, but he punished Prometheus by having him chained to a rock, where every day an eagle came and ate his liver, which grew back again each night.

The tale ends more happily, since eventually Hercules killed the eagle and freed Prometheus.

Like many ancient myths, the legend of the fire giver is fantastic in detail yet remarkably faithful in spirit to the realities of prehistory. Anthropologists who have sifted through the dust and fossilized remains of bygone millennia have found a picture of humankind's discovery of fire that is much less romantic yet startlingly close to the essence of the myth.

The first evidence of man's use of fire dates back roughly half a million years, thousands of centuries before the Greeks began telling their stories. The hero of the fossil record is hardly godlike in appearance. He is *Homo erectus,* an ancestor of ours who lived in Africa, Asia, and possibly Europe during the warm epoch between the second and third glaciations of the Ice Age.

For the past million years, our planet has been subjected to long periods of cold where glaciers have crept from the frozen polar regions and the tops of high mountains to cover much of the world. Most of Europe and much of North America have been covered with ice sheets more than a mile thick. So much of the world's water was frozen into these ice sheets that the very levels of the oceans dropped hundreds of feet in some places.

That period of geologic history is called the Pleistocene epoch, from Greek roots meaning "most recent." Geologically speaking, it is "most recent," a bare million or two years ago since it began; its ending is arbitrarily dated at about 10,000 years ago, when the last great Ice Age glaciation retreated, leaving the world's climate pretty much as it is today.

By "the last great glaciation," I do not mean the *final* advance of the smothering ice sheets necessarily. We may still be in the Ice Age, merely living through a warm spell between the long millennia of cold. No one truly knows for certain; all we can say

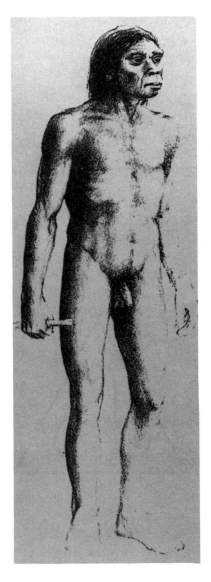

The reality: Scarcely five feet tall, with a brain only two-thirds the size of our own, it was Homo erectus *who tamed fire, about half a million years ago.*

is that about 10,000 years ago the great ice sheets had melted back toward the poles and the high mountaintops. When they might advance again is still a mystery, although geologic evidence has shown that there were earlier ages of ice hundreds of millions of years ago.

Homo erectus, the real Prometheus, existed during one of the warm interglacial periods, about half a million years ago. Scarcely five feet tall, with a skull rather halfway between the shape of an ape's and our own, little *H. erectus* had a cranial capacity of about two-thirds our own. Scarcely the stuff of poets or astronomers, yet enough brainpower to make the most significant discovery in our history.

Anthropologists have dubbed this ancestor of ours with the *erectus* tag because he stood erect. His legs and feet were as good as ours, although smaller. His hands were like ours, with those wonderful opposable thumbs that allow us to make and use tools. He walked, and although he probably did not talk, it is likely that he could communicate with his kind as well as modern chimpanzees do, if not better.

He undoubtedly had two other attributes that we would recognize and admire: curiosity and courage. With them, and his physical gifts, *H. erectus* tamed fire and opened a new era for himself and his descendants.

It was not an easy life in those grassy savannas of half a million years ago. Many *erectus* skulls and bones have been found with the scratches of leopard's teeth gouged into them. Small, slower than the great cats that stalked them, pitifully weak compared with the bears that competed against them for cave shelters and food, our ancestors had to make use of their hands and minds—or die out completely.

In cave sites as far distant as Hungary and China anthropologists have found the ashes and charred sticks and bones, fossilized into stone, that are the remains of campfires lit some half-million years ago.

How did this protoman tame fire? Very likely fire was truly a gift of the heavens: a stroke of blinding lightning that set a tree alight or started a brush fire that swept across some Pleistocene valley. Possibly some foolishly curious *erectus* youngster snatched at the burning branch of a bush or tree. The first result of such bravery was probably a set of burned fingers and a yowl of pain.

But soon enough, *Homo erectus* learned how to handle fire. He took the gift of the heavens, the gift of Prometheus, and started on the road to world dominion.

Fire gave our ancestors tremendous advantages over all the other creatures of the Pleistocene. First, fire frightened away the beasts that slinked through the night and preyed on sleeping animals. While still among the slowest and most poorly armed creatures of the Ice Age landscape, early man became much safer and more secure at night. We still have our fear of the dark, a deep racial memory of the time before we could create our own light and frighten away the predators and imaginary demons that lurk among the shadows. But we are no longer leopard bait.

Second, fire provided the warmth that allowed primitive tribes to expand their habitats and move into the colder parts of the world. We are tropical creatures, at heart. Our own blood temperature tells us that. Even Eskimos heat their igloos to tropical temperatures, despite the Arctic cold beyond their walls of ice. The earliest ancestors of humankind have been found in the warm regions of Africa, Indonesia, and southern Asia. During the enor-

Fire is fundamental to human existence. It provides warmth, protection, the ability to cook food—and a deep emotional resonance. Depicted here is a scene from the film Quest for Fire, *in which a prehistoric tribe discovers the process of making fire.*

mous climatic upheavals of the Pleistocene, fire allowed our ancestors to bring their tropical habitat with them wherever they went,
even up to the glittering face of the towering ice sheets that covered so much of the land. This was of crucial importance during
the millennia-long winters of the Ice Age.

Other creatures had to migrate away from the ice, or grow
shaggy coats of fur, or disappear into extinction. Primitive humans
not only survived the age of ice, they flourished. They spread all
around the world. And when the last glaciation began to melt
away, our ancestors had evolved to the point where they were as
fully human as you and I.

Fire may even have helped in that physical evolution. *H. erectus*
ate his food raw, as every other creature does. *H. sapiens,* you and
I and our Neanderthal and Cro-Magnon ancestors, cook our
food. Cooking not only kills off harmful microbes dwelling in the
food; it also softens the food and makes it easier to eat. We may
argue over the proper amount of cooking for this or that particular
dish, but even a purist who insists on having his spaghetti al dente
would blanch at the thought of eating the pasta raw.

Cooking meant two things to our ancestors. They could spend
much less time chewing and eating, and thereby devote more time
to hunting, exploring, and generally enjoying life. A gorilla must
spend its entire waking day stuffing vegetation into its mouth
merely to keep its impressive body going. A human being can get
just as much nourishment in minutes by eating cooked meats.

Less chewing meant that the apelike muzzle of *H. erectus*
could shrink down to the more human mandible of *H. sapiens.*
And while this physical change was taking place, the braincase
grew commensurately. No one can say if the two developments
depended upon one another. All we know for certain is that as
the jaw got smaller, the braincase enlarged. If you think of the
tip of the nose as a fulcrum point, the whole *Homo* face rotated
over the course of a half-million years, with the jaw sinking away
and the forehead expanding dramatically.

Perhaps the most important consequence of taming fire was
that fire became a source of energy for the human race, the first
energy source any creature found outside of its own cells. Yes,
chlorophyllic plants transform sunlight into carbohydrates, but
that energy generation goes on within the cells of the organism.

With fire, human beings escaped the limitations of biological energy transfer. They went beyond the strength of their muscles, beyond even the strength of other animals they tamed.

Fire started the human race on the road to civilization. With fire our ancestors learned how to make clay vessels and fire-harden wooden spearpoints. They began to smelt metals and generate electricity. The history of civilization can be seen as the chronicle of man's harnessing constantly hotter and hotter fires, more energetic sources of energy. Wood gave way to charcoal, then the age of steam was built upon coal fires. Now oil and natural gas and even the furious heat of the atomic nucleus fuel humankind's civilization.

In the Prometheus myth, the godlike giver of fire was punished for helping man. In reality, we punish ourselves. The waste products of our fires threaten to foul the air we breathe and the water we drink so completely that the survival of our civilization may be at stake. Slowly we are learning to control ourselves. The

Fire led to civilization, industrialization, and pollution. The gift of Prometheus is not without its undesirable side effects.

Stone Age hunter gave no thought to the smoke of his campfire as it wafted into the pristine air. Today we are painfully aware of how the effluents of our smokestacks and the exhausts from our automobiles affect the quality of our lives and our health.

The Gift of Eve

But long before the human race had grown to such numbers and power that we became a threat to the planet's ecological balance, our ancestors made a discovery that influenced the course of life on Earth just as much as the discovery of fire did. More so, in some ways.

It was a discovery that allowed our forebears to harness light in a new way. New to them, that is.

Ten thousand years ago our ancestors made the key discovery that transformed the human race from bands of nomadic hunters into builders of mighty cities. By capturing the energy of light in a new and more efficient way, the population of the human race rose from perhaps 100,000, worldwide, to more than 5 billion today.

This discovery was so profound, so dramatic, that it became enmeshed in the legends of Genesis. It was the discovery that took the human race out of Eden and forced us to "earn thy bread by the sweat of thy brow."

It was the discovery of agriculture.

Ten thousand years ago, give or take a few centuries, the human race consisted of a few scattered nomadic tribes of hunters. It was truly a time of paradise. The last glaciation of the Ice Age had melted away; the ice sheets had receded, opening much of the landmass of Europe, Asia, and North America. Herds of game teemed everywhere, it seemed, and our nomadic ancestors followed them. They were already mighty hunters who had driven several species of big game into extinction.

The word *paradise* is of Persian origin. It means "hunting preserve." Anthropologists point out that for virtually all of the human race's existence, starting with little *Homo erectus* of a half-million years ago, our ancestors have been hunters. Except for the past hundred centuries, a mere eyeblink compared to the

5,000 centuries preceding them, human beings have built their societies, their ethics, their family relationships, around the idea of nomadic hunting.

Even our bodies and minds are shaped for hunting. As Coon puts it in *The Story of Man:*

> One fact about the life of a hunter which all who have lived among such people have noticed is that hunting is fun. Hunters take pleasure in their work. Human beings have been hunters for a long time, and our physiology is adjusted to this kind of life. . . . A man is at his best from the standpoint of fertility if he is away from home a night or two at a time, giving his sperm cells a chance to accumulate. Hunting gives him just these little absences. Hunting exercises the whole body, as few other occupations do. It places a premium on keen eyesight. Farsightedness is an asset. . . . Hunting develops the muscles and tissues of the hands properly, instead of deforming and thickening them as farming and unskilled labor may do.
>
> It also places a premium on the capacity to make quick decisions, to act quickly, and to work in teams. Obedience and leadership can be developed in no better school. Courage is also a necessary component, and that peculiarly human thing, the willingness of a man to die in order to save other members of his group. Man is a creature fashioned around and selected for hunting. Wealthy men who can pick and choose their ways of spending their days like to hunt. . . . A good hunter is always on the alert, always ready to take advantage of unexpected circumstances. He is constantly seeking for new things. In our society, the men who find fun in their work and need no hobbies or vacations are the scientists and research men, including the archeologists and anthropologists, who have carried the hunting spirit into new fields.

Agriculture put an end to the hunting life. Once human beings realized that they could deliberately plant grain and harvest it, thereby assuring a food supply, they gave up hunting and settled down on the soil to become farmers. Not every human tribe gave up hunting. Even today there are nomadic shepherds in parts of Africa and Asia and even primitive hunting peoples in remote jungles. But so successful was this new idea of farming that the agriculturists quickly outnumbered all other kinds of societies. By the time of the invention of writing, which probably took place about 5,000 years ago, agricultural societies such as ancient Egypt numbered their members in the millions.

The tale of the Garden of Eden can be seen as nothing less than a lament over the lost paradise of the hunting life. Adam, who literally named all the animals of Eden (a good hunter must know his prey), turned to picking fruit and, eventually, to growing the grain that would provide his bread. Significantly, the lead in this development was taken by Eve. It was the women who stayed at camp while the men went off hunting, the women who gathered fruits and berries, who undoubtedly began planting the earliest crops.

The myth of the Garden of Eden can be seen as a lament over the change from early humankind's life as nomadic hunters to the settled— and arduous—life of farming.

Very likely those nomadic hunting tribes returned to the same areas year after year, following the migrations of the game herds. It must have been the women who noticed that the same kinds of plants grew in the same places and ripened at the same time of each year. It must have been women, whose total existence revolved around producing babies and nurturing them, who made the connection that led to deliberate planting and reaping. Certainly agricultural societies as recently as Roman times celebrated the fecundity of the earth with human fertility rites. An echo of this sympathetic harmony between human copulation and the success of the year's crops can be found in the maypole festivities of medieval Europe, despite their Christian veneer.

Most thinkers believe that bread was the basis of civilization. They reason that the early tribes settled down on a particular piece of farmland when they realized that they could grow grains that could be turned into nourishing, sustaining bread—the staff of life. I have a suspicion that it was not bread but beer that gave the major impetus to farming. Grains can be fermented into heady alcoholic beverages, and every primitive society has a local version of beer—even those remote hunting tribes that have never taken to agriculture.

Anthropologist Solomon Katz of the University of Pennsylvania believes Neolithic hunter-gatherers discovered beer accidentally while soaking wild wheat and barley in water to make gruel. Natural yeast in the air converted the gruel to beer. The brew not only made Neolithic hunters feel good, it actually contained more nutritional value than unfermented grains.

Beer or bread, most of our ancestors forsook their nomadic hunting ways and became farmers. The human race has never been the same since.

With the invention of farming, human tribes gave up their roaming hunting life and settled down to become "sons of the soil." Nomadic hunting tribes are roughly democratic. No one owns much in the way of property except the tools and clothes he or she carries. Certainly no one owns the land the nomads walk on. The very idea of *owning* the land is totally foreign to hunting tribes. This caused enormous misunderstandings between the American Indians and the Europeans who took their land from them.

Farming is completely different from the nomadic life. The land becomes precious. The man who owns the land also owns the people who work it, or he did until the past century or two. Some people can be wealthy, others poor. Democracy gives way to other forms of rule because one man, or perhaps a few men, must decide when to plant, when to reap. Orders must be relayed to the men and women in the fields. And those orders must be obeyed. I am willing to bet that the whip was not invented until after the invention of agriculture.

In lands where irrigation was vital to the success of agriculture, societies arose that have been dubbed "hydraulic tyrannies." That is, the men who controlled the irrigation system became absolute rulers of the society. In ancient Egypt, the Tigris-Euphrates Valley, and in ancient China such rulers became not only absolute monarchs but gods, literally worshiped by their people.

Farming alters the ecological balance of the land. Grasslands are plowed under, to be planted with seasonal crops. Forests are burned or chopped down to make room for more cropland. Streams are dammed or diverted; irrigation canals are dug.

But despite the wrenching changes and drawbacks that farming brought, once a tribe turned to farming it hardly ever went back to the hunting way of life. The advantages far outweighed the disadvantages, no matter how the tribal storyteller might lament the loss of the old ways in paradise.

Purely and simply, agriculture allowed those first farmers to get far more food out of an acre of ground. In essence, human beings discovered the same trick that life itself had stumbled onto some 4 billion years earlier: the trick of creating foodstuffs out of sunlight, water, soil, and air. Green plants make their own food. With agriculture, clever humans learned how to do the same thing—at one remove. They cultivated the plants that made foodstuffs, and then ate them.

Where it took a team of hunters many days to track down a single deer over an area of several square kilometers, the same number of men turned to farming could produce enough food to feed hundreds of people in much less territory. The crops were used in part as food for humans, in part as food for the animals that humans were now domesticating. Farming forced people to settle in one place. That, in turn, allowed them to domesticate

the pig, chicken, cow, and other animals to the point where many such species no longer exist in the wild; they are dependent on man for their continued existence.

The two key inventions of human prehistory were fire and agriculture. The creation of light at man's behest, and the use of light to grow food crops. No civilization exists without them. The Incas could build a civilization that spanned the Andes without inventing the wheel. The Romans could build aqueducts still in use today without developing steam engines. The British could conquer a world-girdling empire before the development of electricity. But no civilization has ever been created without the use of agriculture, the gift of Eve. And no human tribe, no matter how primitive, exists without the fiery gift of Prometheus.

Agriculture allowed human beings to use the same trick life itself had discovered some 4 billion years earlier: create food out of water, soil, air—and light.

71

7

The Glow of Health

IT'S A LOVELY summer afternoon, and you decide to soak up a few rays out in the privacy of your backyard. Wearing a swimsuit and armed with a bottle of sun block, you stretch out on a reclining beach chair with a copy of Ernest Hemingway's classic novel *A Farewell to Arms*.

Reading Hemingway's terse prose, you soon find that in his story, sun-filled days are associated with love and happiness, while rain brings death:

> . . . and in the fall when the rains came the leaves all fell from the chestnut trees and the branches were bare and the trunks black with rain. . . . At the start of the winter came the permanent rain and with the rain came the cholera.

At the end of the novel, when Lieutenant Henry's lover, Catherine, lies dead in a hospital room:

> But after I had got them out and shut the door and turned off the light it wasn't any good. It was like saying good-by to a statue. After a while I went out and left the hospital and walked back to the hotel in the rain.

Rain and death. Darkness and misery. Over the long eons of humankind's existence on Earth we have come to associate

darkness not only with danger but also with death itself, the ultimate destroyer. We seek the light. Even in the shallowest horror tales, fire is the purifying agent that can kill the monsters. Fire is dangerous but useful. Darkness is fearsome and symbolic of death.

Sunlight exerts powerful influences on our bodies, our minds, our moods, and our health.

This love of light and fear of darkness is not merely a matter of psychological mood or social training. As you lie on your beach chair in the bright afternoon, the light from the Sun is exerting a powerful effect on your body—and your mind. In fact, light has a direct influence on the workings of the innermost part of the brain.

And our bones.

When I was an infant, during the darkest days of the Great Depression of the 1930s, I suffered from the bone disease called rickets. My body could not make enough calcium to keep my bones as strong as they should be. Rickets sufferers can become crippled for life and have grotesquely bowed legs or a malformed skull or chest. The disease also weakens the muscles, leaving the patient even more crippled. Fortunately, mine was a mild case and I eventually grew up with a normal skeleton inside me, although I have suffered ever since from asthma, which may have started when, as a baby, I had barely the strength to lift my rib cage and breathe.

Rickets is caused, in part, by a lack of sunshine. Poor diet also plays a role. An infant living in the narrow streets of south Philadelphia during the dreary winter of 1932–33 got precious little sunshine. And with the Depression, my diet was not as nutritious as it needed to be. My first Christmas, the family larder contained nothing but a can of baked beans; the family treasury was down to one dime.

Rickets is a vitamin-deficiency disease. In this case the vitamin is D. The prime source of vitamin D comes from the ultraviolet light in sunshine, the same UV that tans our skins. The ultraviolet generates vitamin D from chemicals in the skin called *sterols*. One of these sterols, ergosterol, is obtained from vegetable oils in the diet. Thus a child who gets little sunshine and few vegetables can be susceptible to rickets.

My memory of the cure is worse than my memory of the disease. For years and years my mother rammed a large spoonful of cod-liver oil down my throat every morning, rain or shine. Cod-liver oil is rich in vitamin D. Its taste is something else.

As you lie out in the backyard sun, you should understand that ultraviolet light is not an unmixed blessing. Only a tiny amount of it gets through the ozone layer high in the atmosphere and reaches the ground. However, the ozone layer is thinning out at an alarming rate. Each spring for the past several years, a large and growing gap in the layer has appeared over Antarctica.

Many scientists are convinced that the chemicals released from aerosol sprays are destroying the ozone high up in the atmosphere. International efforts are under way to stop the manufacture of such canned aerosols. Some scientists believe that the ozone "hole" is a natural phenomenon, tied to long-term variations in the Sun's radiation, and will disappear when the Sun returns to "normal." In either event, a reduction in the ozone layer spells deadly danger for the human race—and most other forms of life as well.

UV helps to produce vitamin D. But it also causes skin cancer. Too little UV and the result is not only rickets but women whose pelvic bones are so deformed that they die in childbirth. If our ancestors had not received enough ultraviolet from sunshine, the human race would have become extinct, victim of overwhelming rates of infant and maternal mortalities.

Yet too much UV produces skin carcinomas and melanomas, cancers that are equally dangerous. Untreated skin cancers can spread through the body and lead to agonizing death. The suntan lotion you smear on your exposed skin will help to block the UV rays, but if the ozone layer continues to be depleted, sunbathing will eventually become as hazardous as cigarette smoking.

Sun-blocking lotion or not, after half an hour you go back inside your house and sit in front of your computer. You turn it on (receiving a mild dose of UV and low-energy X-rays square in the face) and call up on the display screen a map of the Eurasian landmass. By pressing a key you get the screen to show where the Caucasian peoples—the so-called white race—have made their homes. From the North Cape of Norway, more than 300 miles above the Arctic Circle, to steaming Sri Lanka, a scant ten degrees north of the equator, the Caucasians have lived since before the invention of writing.

Now you press keys that show the shadings of skin, hair, and eye coloring of Caucasians in each region. Caucasian coloring varies from the pale, blond, blue-eyed Nordics to the virtually black, brown-eyed, black-haired Tamils of Sri Lanka. The white race can be many colors: pink, cream, tan, brown, or black, with all sorts of shadings in between.

Finally, you press the key that shows the number of cloudy days per year in each region. Not surprisingly, the paler a Caucasian's coloring, the cloudier the climate he or she was born to. The darkest Caucasians have inhabited the lands where the sun burns most fiercely.

Skin color among Caucasians varies from very dark to very fair, primarily as an adaptation to the amount of sunlight available in the region the group inhabits.

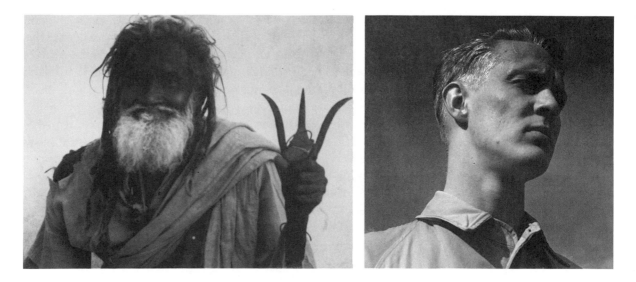

For Caucasians, skin, hair, and eye coloring are adaptations to the amount of sunlight available and particularly the amount of ultraviolet in the sunlight. Nordics need to take in as much UV as they can because so little penetrates to the ground in their high-latitude cloudy lands. The darker peoples of the great Indian subcontinent need to protect themselves against too much UV, and the darkness of their skin is the result.

Any discussion of race or skin color stirs passionate debate even among unbiased academics. The evidence seems clear that the various races of humankind made certain areas of the globe their homelands long ages ago, and among the adaptations to the climates in those homelands that developed was the coloring of their skin, hair, and eyes. These were certainly not conscious adaptations, and biologists can argue long and loud over whether the different peoples moved to regions that suited their existing coloration or whether they adapted to the regions after they settled in them.

To us, the point is this: Humankind has learned to live with ultraviolet, adapting our coloration to provide enough UV for the necessary production of vitamin D, but to screen out the excess UV that can cause skin cancers.

The pigment that handles ultraviolet is melanin, and it is present in every human being in our skin, our hair, and the irises of our eyes. Fair-skinned people have comparatively little melanin, but that which they do possess darkens the skin upon exposure to ultraviolet light. They acquire a tan, a protective coloration against too much ultraviolet. In dark-skinned peoples greater amounts of melanin are present in the skin, able to block virtually all of the incoming UV and thus protect the body from the harmful effects of excessive ultraviolet radiation.

In addition to producing vitamin D, light can have other beneficial effects, and the medical profession is moving cautiously into the field of *phototherapy.*

The idea of using light to treat disease is not all that new. In 1876, Augustus J. Pleasanton, a retired Civil War general, published *The Influence of the Blue Ray of Sunlight and the Blue Color of the Sky, in Developing Animal and Vegetable Life, in Arresting Disease, and in Restoring Health in Acute and Chronic Disorders to Human and Domestic Animals.* His book made such broad claims for the beneficial effects of light that it sounded somewhat like the pitch a snake-oil salesman might make from the back of his wagon: "Good for man or beast!" His book was bound in blue and printed in blue ink.

Several other books about treating disease with light were written by men of varied (and sometimes doubtful) scientific training, and for a while there was a fairly serious effort at what has come to be called *chromopathic medicine*. What little good chromopathic practitioners accomplished was drowned in the flood of quacks and frauds who made money by shining light on sick people to no medical avail.

The reason most light therapies did not work is quite simple. The human body does not absorb light very well. Shining light on a sick person is rather like shooting bullets at Superman. The stuff just bounces off.

Most of it. But not all.

There is one compound that the human body manufactures in plentiful quantities that does absorb light and does undergo a photochemical change. That compound is called bilirubin. In a normal person bilirubin is metabolized by the body as quickly as it is produced and is excreted away. But newborn infants, especially those born prematurely, are susceptible to neonatal jaundice, usually caused by the inability of the preemie's incomplete liver to break down bilirubin and get rid of it.

High concentrations of bilirubin in the blood of a newborn can cause the jaundicelike brown coloring of the skin, brain damage, mental retardation, and even death. Neonatal jaundice is usually treated by transfusions, replacing the bilirubin-loaded blood with fresh blood, or by drugs.

Or by blue light.

Blue light is absorbed by bilirubin, and the energy of the light causes the compound to break down into chemicals that the newborn baby can easily excrete. Phototherapy, to date, has shown a good success rate against neonatal jaundice, without the harmful side effects that drugs might cause or the dangers of wholesale blood transfusions.

Still, medical doctors are moving cautiously. Clamping a set of blazing blue lights over a preemie's crib calls for certain precautions. The infant's sensitive eyes must be protected, and the

PLATE 1
One of the oldest forms of life, the microscopic blue-green algae began to harness the energy of sunlight more than 3 billion years ago and are still doing it today.

PLATE 2
Satellite imagery of the "hole" in the ozone layer that appears each spring over Antarctica. Air pollution appears to be a major contributor to this weakening of the ozone layer.

PLATE 3
Deep beneath the ocean's surface, marine creatures generate luminescent light through chemical reactions in specialized cells of their bodies.

PLATE 4
The colors of the iris in human beings range from deep brown to pale blue, and even pink in albinos.

PLATE 5
On Earth, the sky appears blue because the blue wavelengths of sunlight are scattered by the gas molecules in the air more than the longer, redder wavelengths. On Mars, the sky appears pink because there is so much dust in the air scattering the red wavelengths of sunlight.

(Captions continue on page 79.)

PLATE 1

PLATE 2

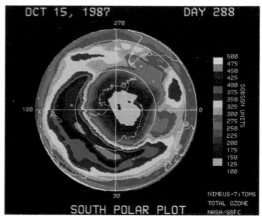

OCT 15, 1987 DAY 288

500
475
450
425
400
375
350
325
300
275
250
225
200
175
150
125
100

DOBSON UNITS

NIMBUS-7:TOMS
TOTAL OZONE
NASA/GSFC

SOUTH POLAR PLOT

PLATE 3

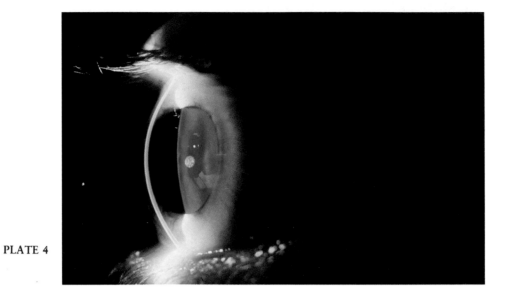

PLATE 4

PLATE 5

PLATE 6

PLATE 7

PLATE 8

PLATE 9

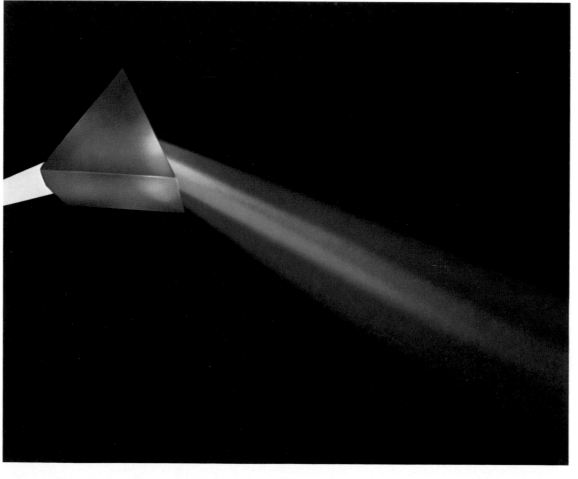

cribs of other infants in the ward must also be shielded against the glaringly bright light. Happily, the chemical by-products of the bilirubin that has been broken down by phototherapy have been found to be nontoxic.

Laboratory studies have shown, however, that intense blue light can modify the DNA at the heart of living cells and cause other chemical changes in vitamins, amino acids, and proteins. While no deleterious effects have been found in the babies saved from neonatal jaundice by blue light treatments, researchers caution that phototherapy should be used carefully and that more work must be done on refining the treatment.

While it is known that ultraviolet and even visible light can cause damage to living tissue, light can also be a curative agent in cases of psoriasis, vitiligo, and even certain forms of cancer.

Psoriasis and vitiligo are skin disorders that produce ugly, itching sores. In fact, *psoriasis* is a word of Greek derivation that means "itching." While not terribly threatening diseases, they are highly uncomfortable and socially embarrassing. (Remember the skin-ointment advertisement that spoke of "the heartbreak of psoriasis"?)

It is a standing joke in the medical profession that if you really want to get rich, you should become a dermatologist: Your patients won't die of the diseases they come to see you about, and they will never be cured. There are no cures for psoriasis and vitiligo, merely treatments that alleviate the symptoms and shrink or cover up the ugly scaly patches on the poor victim's skin. Taber's *Cyclopedic Medical Dictionary* states that the causes of these maladies are unknown, and the treatments recommended merely "give comfort to the patient as well as help to control the disease."

Cancer is another matter altogether, the second-highest cause of death in the United States, which attacks nearly a million Americans each year and kills close to half a million annually.

Psoriasis, vitiligo, and—most importantly—at least two forms of cancer can be treated by light therapy.

Phototherapy for these diseases hinges on the discovery that certain compounds called *psoralens* can be activated in the body by ultraviolet light to break down the DNA in the diseased cells. Psoralens occur naturally in a wide variety of fruits and vegetables, including figs and limes. Moreover, a synthetic form of psoralen has been developed, 8-methoxypsoralen (8-MOP), which can be administered orally and is much more concentrated than the psoralens found in nature. (How many figs can you eat?)

Once in the body, psoralens are quite innocuous under ordinary circumstances. But when exposed to intense ultraviolet light, they become active molecular surgeons, microscopically snipping the chemical links that hold together the DNA molecules of the cells involved in psoriasis, vitiligo, and two of the deadliest forms of cancer: leukemia and T-cell lymphoma.

To fight these forms of cancer, medical researchers give the patient 8-MOP and then route the patient's blood out of the body through plastic tubing. The blood is exposed to intense UV light and then pumped back into the body. The cells of lymphomatic leukemia and T-cell cancer are destroyed by the UV-activated 8-MOP. In psoriasis and vitiligo cases, the UV light is applied directly to the area of the skin affected.

Again, phototherapy is a comparatively new form of treatment. Research is continuing to determine the limits of its efficacy and its side effects. But it is heartening to know that light, albeit a form that is invisible to our eyes, is being used to strike down the dark scourge of cancer.

Phototherapy is being studied, and used, for several other kinds of diseases, particularly skin diseases. Research is under way to see how UV and visible light might affect the immune system.

There is one effect of light that strikes many people almost every day. Perhaps it happened to you when you first went out for your afternoon stretch in the sun. You sneezed.

In biomedical parlance, you were a victim of the *photic sneeze reflex.* Just as a bright light triggers a reflex in the eye, contracting the iris before harmful amounts of light can damage the retina, bright light also can trigger the sneeze reflex. In his book *The Body in Time,* biologist Kenneth Jon Rose says, "While no one knows the exact mechanism for this effect [the photic sneeze reflex], it is believed that light, through a torturous network of nerves from the visual system of the brain, activates receptors in the nose triggering a sneeze. The tight relationship between a sneeze and the eyes can be demonstrated by its behavior: Bright light triggers blinking and tearing reflexes. So does a sneeze."

Researchers may not know why bright light causes some people to sneeze, but they have a name for this trait. They call it ACHOO, for Autosomal dominant compelling Helio-Ophthalmic Outburst syndrome.

Well, you've had your half-hour in the sun, with or without sneeze. Now, as you sit in front of the glowing computer screen, you begin to wonder about the health effects of the many sources of light inside your home: incandescent light bulbs, fluorescent lamps, computer and TV screens.

There are two major differences between artificial light and natural sunlight. First, artificial light does not come in all the colors of the rainbow, as natural light does. Second, compared with that glorious daystar of ours, artificial lighting is *dim*.

Sunlight is called white light by the physicists. By this they mean that sunlight is composed of a mixture of every color of the spectrum, from deepest violet to hottest red, plus colors that are invisible to us, ultraviolet and infrared. Actually the Sun puts out an even broader spectrum of radiation that includes more energetic forms of ultraviolet, X-rays, gamma rays, and even radio waves. All of these nonvisible wavelengths are blocked by the atmosphere, except for some of the radio waves.

Moreover, the atmosphere scatters and absorbs some wavelengths of light more than others. This is why the sky looks blue. The shorter blue wavelengths of sunlight are scattered by the molecules of gas in the atmosphere much more than the longer, redder wavelengths. The result: blue sky.

At dusk, as the light of the setting Sun struggles through a longer path in the atmosphere than when it is directly overhead, the dust and other pollutants in the air scatter the reddish wavelengths, producing the brilliant hues of sunset. As a Texan friend of mine told me while we admired a beautiful sunset in Houston: "Brought to you through the courtesy of industrial waste."

On the planet Mars the sky appears pink, mainly because there is so much reddish iron-ore dust in the thin atmosphere of that waterless world that the red wavelengths of light are scattered more effectively than any other color. (See Plate 5 following page 78.)

A brief word on wavelengths. The wavelengths of light and the other radiation emitted by the Sun run the gamut from miles long, for some of the radio waves, to submicroscopic. This range of wavelengths is called a *spectrum*. You can see the spectrum of visible light whenever a rainbow is in the sky; a rainbow is nature's way of producing a spectrum.

Visible light lies in the range of about 400 to 700 *nanometers*. A nanometer is one-billionth of a meter (10^{-9} m), so small that

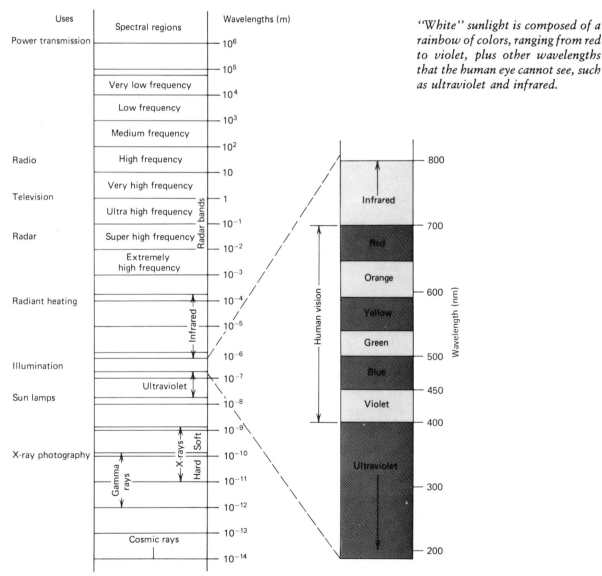

"White" sunlight is composed of a rainbow of colors, ranging from red to violet, plus other wavelengths that the human eye cannot see, such as ultraviolet and infrared.

it would take 10 million nanometers to span the width of your little finger. In English units, a nanometer is slightly less than four ten-billionths of an inch (3.937×10^{-10} inch). For this discussion we will stick to the metric units. Scientists and engineers also sometimes refer to wavelengths in terms of *micrometers,* which are one-millionth of a meter long. In other words, one micrometer equals a thousand nanometers.*

No matter which unit used, visible light comes in extremely small wavelengths. But then good things usually come in small packages, it's been said.

The sunlight that reaches the ground falls between wavelengths of roughly 200 to 3,200 nanometers. The shorter wavelengths are, of course, ultraviolet light, while the longer are infrared. The light that our eyes can ordinarily see, as we said, lies roughly between 400 and 700 nm. Those numbers are not precise because some people can see a wider range of colors than others. We are dealing with human perceptions here, not mathematical precision, and individual circumstances vary. There are people who can see ultraviolet—although they can never explain what it looks like to those of us who can't. The widest range of wavelengths that humans can see is apparently from around 380 to roughly 800 nm.

In addition to its broad spectrum, sunlight is also very bright. Lighting engineers measure brightness levels in terms of the unit called *lux,* which is roughly equivalent to an energy of 18 watts per square foot. Measurements made in various temperate-latitude locations show that sunlight reaching the ground varies in brightness from an average of nearly 60,000 lux in midwinter to almost 114,000 lux in midsummer.

(For those interested in solar energy, this means between 29 and 57 watts per square meter, a bonanza of free energy falling out of the sky. An average-sized suburban house, for example, has a roof area of roughly 2,000 square feet. That means that between 58 and 114 kilowatts of solar energy strike that roof whenever the Sun shines! In my home in Connecticut, that energy heats our water, as well as the southward-facing extension we put on the house.

*A smaller unit often used in dealing with the wavelengths of visible light is the *Angstrom,* named after Swedish physicist Anders Jonas Ångstrom (1814–74). The Ångstrom (Å) equals 10^{-10} m, or 0.1 nanometer.

Sunlight is much brighter than even the most powerful artificial lighting.

Even in midwinter our sun room is tropically warm as long as the Sun is shining. When solar cells become cheap enough we plan to generate our own electricity from sunlight, just as spacecraft do.)

Artificial lighting does not come even close to the broad spectrum of sunlight. Nor its brightness.

Old-fashioned incandescent lamps, the kind that create light by heating a filament until it glows, give the broadest spectral range. Incandescent lamps typically glow in wavelengths from roughly 250 to beyond 900 nanometers, but most of their brightness falls in the yellow-red end of the spectrum, from 650 nanometers up.

Fluorescent lamps emit light in much narrower wavelengths, since they glow from the light given off by specific atomic elements, usually either mercury or sodium. Most of the fluorescents used in homes and offices are mercury lamps, which give off a bluish light that can make skin look ghastly and all colors seem eerily strange.

When I was a magazine editor, my Manhattan office was illuminated with mercury-vapor fluorescents. Artists who worked as illustrators for the magazine constantly complained that the office lighting distorted the colors of their paintings. Sure enough, if we went into an area lit by old-fashioned incandescents or, better yet, went out into the sunlight, the colors looked much better.

It is no secret that lovers love candlelight. Not only is it more romantic than electric lighting, the soft yellowish glow of candlelight also makes skin tones appear warmer and smoother. I have also noticed that candlelight produces a lovely sparkle in your dinner partner's eyes, well worth any difficulties in identifying the food on your plate that might be caused by the reduced light level.

Fluorescent lights flicker. Electrical current stimulates the gas atoms inside them to glow; the current switches back and forth 60 times per second. This is too fast to bother most people. But when your field of vision places a fluorescent lamp "in the corner of the eye," where the very sensitive rods are looking at it, you probably notice the flicker. Stare straight at the lamp, so that the cone-rich foveae are pointed at it, and the flicker disappears. Look slightly away again, and the annoying flicker can again be seen.

The glass envelopes around each kind of lamp cut off virtually all the ultraviolet, but you can feel the infrared pouring out of an incandescent lamp. After all, the gadget only emits light because the filament inside it is heated by electricity until it glows.

Fluorescents are much cooler; the electrical current flowing in a fluorescent lamp is exciting individual atoms of mercury or sodium vapor to emit light without heat.

Compared with the 60,000 to 114,000 lux brightness of sunlight, artificial lighting is quite dim. Fifty lux is considered bright enough for areas in business offices in which computer display terminals are being used. To read penciled handwriting, lighting engineers say 2,000 lux is sufficient. Most business offices are lit at a brightness of somewhere between 200 to 1,000 lux—the brightness typical of twilight outdoors!

This may explain a phenomenon recently discovered by researchers at the City College of the City University of New York. In 1987 a research team led by Josh Wallman came to the conclusion that reading can lead to myopia—near-sightedness. Studies done on Eskimos who recently were subjected to compulsory education, and experiments performed with chicks, led Wallman's team to conclude that the task of reading stimulates the foveal region of the retina but not the broader area surrounding the fovea, and this leads to myopia.

Considering the light levels under which we do most of our reading, however, an alternative explanation might be that people become myopic because they are attempting to read in conditions that are too dim.

More and more people are watching television and computer screens every day. The cathode-ray tubes that form these screens emit not only visible light but also a good deal of ultraviolet and even "soft" (i.e., relatively low energy) X-rays. Distance lends some safety. The X-rays and most of the UV are absorbed by a couple of feet of air. Any kind of eyeglasses, even those with nothing but windowpane glass in them, will also block out the UV. Increasingly, opticians are recommending lightly tinted eyeglasses for persons who work all day at computer terminals.

Indoor lighting is very different from the natural sunlight that was humankind's only illumination, except for fire, until little more than a century ago. Yet even though we may spend most of our time indoors under artificial lighting, the natural rhythms of day and night, season following season, have profound effects upon all life on Earth—including humans. These effects come about through a sort of "third eye" in the human visual system, an optical tract that "sees" without vision.

8

Light and Lust

IN WHAT must surely be one of the most erotic poems in the English language, *The Rape of Lucrece,* the evil Sextus Tarquinius extinguishes the torch he has carried into Lucrece's bedroom before he attacks her because, in Shakespeare's words, "light and lust are deadly enemies."

For centuries critics and admirers of Shakespeare have read all sorts of meanings into his works. Psychologists have pored over *Hamlet* and political science majors have written theses about *Coriolanus.* Shakespeare has been given credit for profound insights into the human heart, credit he richly deserves.

But Master William was mistaken in his belief that light and lust are enemies. He understood human behavior well enough, but he did not understand the workings of the pineal gland deep inside the human brain.

How could he? No one understood the functions of the pineal until centuries later. Wonderful observer of humankind that he was, Shakespeare realized that most people prefer to make love at night, and usually in the dark. His poetry reflected his observation of human behavior, not a knowledge of neurophysiology.

The question is, *why* do people prefer to make love at night, in the dark, often with their eyes closed?

Believe it or not, the answer lies partly in the fact that our planet has a tilted axis. The axis of Earth's rotation, the imaginary line that runs from North Pole to South Pole, is not perpendicular to the plane of our orbit around the Sun. The axis is tilted by 23.5

degrees. This means that, as the Earth swings around the Sun in its yearly orbit, different parts of the world receive differing amounts of daylight at different times of the year.

We have seasons.

One of the consequences of having seasons (or one of its causes, depending on how you look at it) is that the days of our years are not all the same length. In summer the days are long and lazy. In winter they are short and cold. When I lived in Boston, it was totally dark by 4 P.M. in midwinter. Friends of mine who live in Norway take part each June in the festivities of the Midnight Sun, the day when the Sun never touches the horizon at all.

"Light and lust are deadly enemies":
Lucrece and Sextus Tarquinius.
Shakespeare's poetry was magnificent; his knowledge of physiology somewhat lacking.

The changing length of the day governs an enormous range of behavior in both plants and animals. And that includes the human animal. Everywhere on Earth, with the exception of some tropical species that live in areas where variations in the day's length are hardly noticeable, life cycles and day-to-day life-styles obey the rhythm of Earth's changing seasons.

In New England the most obvious example of seasonal behavior is the glorious display of colors that the trees put on each autumn. (See Plate 6 following page 78.) The whole landscape turns brilliant red, orange, and gold. It's as if nature is saying, "Okay, I know winter's on the way. Here's something to make up for it; here's some beauty for you to marvel at before the snow covers everything."

It was not until 1920 that biologists figured out that most plants will not flower until the day reaches a certain length. Some plants are autumn bloomers; they will not flower until the daylight hours become short enough. Others are spring bloomers; they will not flower until the daylight hours become long enough. Then there's the third kind, which flower about the same regardless of day length.

The biologists call this phenomenon *photoperiodism,* and the length of day that triggers blossoming is called the *critical day length.*

Even the parts of plants that grow underground—bulbs, tubers, roots, and such—are affected by the critical day length. Obviously, the plant's leaves sense the light and produce hormones that control the rest of the plant's growth, maturation, and even its coloring.

Just as Shakespeare said, it is the absence of light, rather than light itself, that governs the flowering and growth of plants. You can prove this for yourself. Take a long-day plant, any of the many springtime bloomers, and keep it indoors during the winter. It will live and even grow, but it will not make flowers as long as the nights are long. But sneak in on it at night and turn on the lights for an hour or so. The plant will "think" that the nights have grown short and will soon produce flowers, even though the time it spends in light is still well under the length it would be in springtime.

Chrysanthemums are highly seasonal flowers that bloom naturally in the autumn, when nights grow long. But they can be grown in hothouses all year-round by controlling the critical length of darkness with artificial lights. Plants can be kept from blooming until they reach the size the grower wants or until a certain calendar date is reached.

In long-day plants flowering is promoted by the red light portion of the Sun's spectrum, yet for short-day plants red light tends to prevent flowering. Blue light can also promote or suppress flowering, although it is much less effective than red.

Animals are seasonal, too. Even the sophisticated human beings who work in the publishing industry somehow are inclined to bring out new books for the "spring list" or the "fall list." And cultures native to the temperate zone all celebrate a festival of joy a few days after the shortest day of the year. To the Romans it was Saturnalia, the holiday that ended the year. To the ancient Druids of Britain it was the feast of the "turning of the Sun." To their Christian descendants it is Christmas.

The most spectacular example of seasonal behavior among animals is the migration of birds. (See Plate 7 following page 78.) Some species, like the common bobolink, travel thousands of miles from their summer habitat to their winter grounds each autumn, then fly back again in the spring. The white-rumped sandpiper migrates from the top of the world, in the Canadian Arctic, to the bottom, the lower tip of South America called Tierra del Fuego, a distance of more than 8,000 miles each way. The American golden plover flies each autumn from Labrador to the coast of Brazil, then to the Argentine pampas for the winter. It returns in the spring by way of Central America and the Mississippi Valley to the tundras of Alaska and northern Canada, an enormous loop across half the Western Hemisphere.

The long-distance champion is the Arctic tern, which migrates from northern Canada, Alaska, and Greenland to Antarctica!

Ground-dwelling animals migrate seasonally also, always moving to where the good grazing grounds and water holes will be.

Some animals hibernate away the winter. Fishes and frogs can hibernate for months in ice-covered waters where their body temperatures sink down close to freezing. The relatively few species of snakes and other reptiles that live in northern climates go into hibernation for the winter. Bats hibernate. So do squirrels and other small ground-dwelling mammals. And, of course, so do bears.

In the tropics, many species *estivate;* that is, they sleep away the hottest, driest season of the year. In the western deserts of the United States, several species of squirrel go dormant in August and early September. Some gerbils of Africa and Asia also estivate. If you have ever been in Paris during August, you would swear that Frenchmen estivate also.

Actually, human beings are active all year long. (Parisians spend August on the Riviera.) Humans are also sexually active all year long. Like such primate cousins as the gorilla and the chimpanzee, we humans do not have a specific mating season. Rather, for us it is always the mating season.

The key to our seasonal behavior (or lack of it) lies in a tiny knob of tissue buried deep inside our brain. In a very real sense, we have a "third eye" in the middle of our brain that "sees" without vision.

It is called the pineal gland, and it rests beneath the cerebrum, down almost on the brain stem, between the thalamus and the midbrain structures of the limbic system. In other animals the pineal actually contains photoreceptor cells that respond to light very much the way that the primitive, nonfocusing eye of the flatworm does.

The tuatara, of New Zealand, has a functioning third eye in the top of its head.

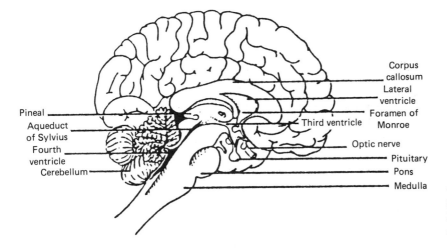

Corpus
callosum
Lateral
ventricle
Foramen of
Monroe
Third ventricle
Pineal
Aqueduct
of Sylvius
Fourth
ventricle
Cerebellum
Optic nerve
Pituitary
Pons
Medulla

Deep inside the human brain, the pineal gland receives information from the eyes that regulates the brain's internal clock.

The tuatara is a lizardlike reptile native to New Zealand. It grows to a length of about two feet and anatomically appears to be a very old form of life, far older than today's reptiles, older probably even than the dinosaurs. Its teeth, for example, are fixed to the edges of its jaws, a very primitive arrangement that modern lizards have "outgrown." The tuatara has a functioning third eye on the top of its bony head. It does not produce true vision the way the two other eyes do, but this pineal eye can sense light and brightness. Biologists believe the pineal eye serves to indicate when the sun is getting so bright that the reptile should get to shelter underground.

In human beings, the pineal structure does not sense light, but it does receive nerve signals that originate in the retinas of the eyes. In effect, the brain has two optical systems; the one we are familiar with, for vision, and another one that stimulates the pineal gland.

The pineal gland secretes a hormone called *melatonin,* which, among other things, inhibits the gonads from secreting their hormones: testosterone in males, estrogen in females. Like the processes that control flowering in plants, the pineal is stimulated to produce melatonin by darkness, not light. Since melatonin inhibits the production of sex hormones (it inhibits ovulation, as well), Shakespeare was wrong when he said that light is the enemy of lust.

What about another English poet, Alfred Tennyson, who observed, "In the spring a young man's fancy lightly turns to thoughts of love"?

Humans do not have a particular mating season, although studies of testosterone levels in human males have shown that there are indeed seasonal and even hourly variations in the amount of testosterone found in a young man's blood. The hormone's highest levels were found to exist in October, the lowest in May. However, testosterone levels varied during the hours of the day; they were highest in the morning, about 8:00 A.M., during the spring, and in early afternoon in autumn. Male sexual activity followed the testosterone levels fairly closely: the more testosterone, the more activity.

So although a young man may think of love in the spring, he has a higher amount of testosterone in his blood in the fall. The actual differences in testosterone level are certainly not enough, though, to produce true seasonal behavior in the human animal. Spring, fall, winter, and summer there is enough passion in the blood to keep the mating game going constantly.

But before the mating game can begin, the body must mature into adulthood. The pineal plays an important role in regulating the onset of puberty, when a lovable child gradually turns into a teenaged monster and—if parents and child survive the ordeal— eventually becomes an adult. *Something* inside the brain counts the months and years, patiently waits until the body has reached the proper growth, and then triggers a flood of hormones that turn girls into potential *Playboy* centerfolds and makes boys want to buy the magazine.

The pineal influences the activity of the pituitary gland, which has been called the body's master gland. Situated below the thalamus, at the end of the hypothalamus, the pituitary is an endocrine gland that secretes hormones that control the body's growth and development and many of its metabolic activities. It is the hypothalamus and pituitary that govern the onset of puberty. Some researchers believe that the pineal can be thought of as one of several alarm clocks within the brain that tell the pituitary when it's time to wake up. When that happens, boys' voices deepen and they start to grow hair on the body and face.

The ritual of shaving begins. Girls begin to develop the anatomical features necessary for bearing and nurturing babies.

It may also be that the pineal gives another signal to the pituitary much later in life. The final signal. The signal that begins the process of physical collapse that ends in death. Biologists are uncertain about this, but some of them believe that the body has a built-in time limit, that no matter how healthy or active a person may be, there is a final limit to the number of days the body will remain alive. It dies of old age because the pineal tells the pituitary and other parts of the brain that it's time to stop.

It's rather scary to think that there is a tiny knot of cells deep inside your brain that is patiently counting the days of your life. Will biologists one day be able to "reset" this internal clock and extend our life spans? Will it be possible eventually to tickle the pituitary into making us more youthful?

The pineal gland recognizes not only seasonal variations in light and darkness. Daily variations are important also and produce rhythms in the body's chemistry that are called *circadian* (about a day). Circadian rhythms affect our growth, body temperature, the way we sleep, eat, work, and live our daily lives.

The daily pace of life is so much a part of us that we take it for granted, except to complain when things get too hectic—or when we are forced to change our circadian rhythm, as when we move from day work to night shift or stay up all night partying or on an uncomfortable overnight flight. Jet lag is a common form of circadian disruption that hits travelers when they fly across many time zones: Your wristwatch tells you it's time for lunch, but the clock inside your body is hollering that it's the middle of the goddamned night!

There *is* a clock inside your brain. Maybe two of them. Maybe even more. Neurobiologists call them *endogenous pacemakers*, which is a specialist's name for an internal clock. Deep inside the brain the pacemaker(s) is(are) counting time, possibly regulated by the pineal's ability to sense the coming and going of light and darkness in the world outside.

Ahah! you say. Maybe the internal clock would work nicely for primitive people (or campers) who live outdoors with the natural rhythm of daylight and darkness. But we civilized types live indoors with electrical lighting. How can the endogenous pacemaker regulate our circadian rhythms when we are exposed to light at virtually all hours, when we can control the hours of light and banish darkness at the touch of a switch?

If you ask that question I must first congratulate you on your command of the neurobiologist's specialized language. But then I must remind you that most artificial lighting is so low in intensity that it fails miserably to impress the pineal and its associated pacemaker. As far as our internal clocks are concerned, artificial lighting is out of sight and out of mind; sunlight, or its absence, is what counts.

We have been trained by eons of exposure to a roughly 24-hour cycle. Infants are apparently entrained to the 24-hour cycle of their mothers while still in the womb. However, when humans are placed in environments where sunlight does not penetrate, such as deep caves or tightly closed experimental facilities, their circadian rhythms tend to drift somewhat. Experimental studies with such subjects have shown that their body rhythms do not go totally wild but rather quickly settle down to some value relatively close to 24 hours and typically fall within the range of 22 to 28 hours per cycle.

Deprived of its external cues of daylight and darkness, the endogenous pacemaker may drift away from a 24-hour cycle, but it does not drift very far.

This may explain a phenomenon we have all experienced: the Monday morning blahs. If the endogenous pacemaker's "natural" rhythm is not exactly 24 hours, it must be reset every day by the light-sensitive pineal, so that we fall into a 24-hour routine each day. On weekends many times we tend to sleep later than usual. We keep our bedrooms darkened so that we can get that extra hour or so of shut-eye. Fine. But on Monday morning we must readjust to the workaday world and its early-rising schedule. Hence, blue Monday.

When humans begin to live on other planets it may evolve that their circadian rhythms will alter to suit the local "day." This will not be a problem for some time to come, though. Not because people will not be living on the Moon or Mars within another generation, but because living quarters on the airless Moon will undoubtedly be underground, where the artificial lighting will be keyed to a 24-hour cycle, while Mars has a roughly 24-hour day, much as our own.* Other space habitats, on more distant and exotic worlds, are another question.

The proof that our internal clock with its circadian rhythms is governed by light comes from the fact that when a nerve tract from the retina is severed, cutting off the signals from the eyes to the pacemaker, circadian rhythms go haywire—even though the optic tract is left intact and the visual system still functions normally. In experiments with animals where the pacemaker has been put out of action, the internal clock no longer runs on a day-night cycle. The animal can still see; its *visual* system is intact. Its body rhythms will run on a cycle that is close to 24 hours, but its activities will be independent of the day-night cycle. A nocturnal rat, for example, will be active in either daylight or night; the amount of light present no longer makes any difference to it.

Our internal pacemaker depends on "seeing" the daily cycle of daylight and darkness; the differences in length of daylight as opposed to darkness trigger the seasonal changes in our body chemistry.

It is fairly easy to understand how the pineal gland, which receives nerve signals from the eyes, can sense the day-night cycle. Less easily understood is how the brain's internal pacemaker counts the hours of darkness in each daily day-night cycle and arrives at the conclusion that spring or autumn has arrived. For this reason, most neurobiologists believe that there are actually two or more pacemakers in the brain.

One of them, just about every researcher agrees, is the *suprachiasmatic nucleus* of the hypothalamus, a tiny region that lies just below the thalamus, deep in the central region of the brain. Researchers, who are just as unhappy with long unpronounceable words as thee and me, abbreviate it to SCN. The SCN is connected to the retinas by two nerve tracts, so it certainly receives optical information. It also lies close to the pineal gland and is undoubtedly connected intimately with it, since both the SCN and the pineal are vitally involved in the pacemaker function.

---

*To be exact, the Earth revolves on its axis in 23 hours, 56 minutes, 4.09 seconds; Mars, in 24 hours, 37 minutes, 22.7 seconds. Close enough for our endogenous pacemakers to easily adapt. The Moon's "day" is exactly the same length as its monthly revolution around the Earth: 27.322 days.

No other pacemakers have yet been identified. Perhaps they do not exist at all, and the SCN does the entire job together with the pineal. However, most researchers reject this idea and believe that there are at least two pacemakers.

Studies have shown that the SCN responds to the presence or absence of light. It acts on brightness, not visual imagery. Like Paul Revere watching the bell tower of Boston's Old North Church, the SCN is looking for a light, not pictures. Either the SCN itself or the yet-unidentified other pacemaker acts on the *length* of brightness and darkness.

We can begin to understand, now, the phenomenon known as cabin fever, or the winter blues. In climates with long winters, many people experience a seasonal bout of depression. The cloudy dull days and long cold nights can make life dreary, dull, and desperately unhappy. Melatonin is secreted by the pineal in the dark, and melatonin in all probability has an overall depressive effect.

A research team at Oregon Health Sciences University led by Alfred J. Lewy showed that a form of phototherapy can alleviate winter depression. Lewy's team exposed eight volunteer patients who had exhibited wintertime depression to bright light (2,500 lux) early in the morning. They found that the patients' production of melatonin was altered, and their depressions significantly improved after a week of such treatment. The bright light in the morning apparently shifted the patients' melatonin production cycle enough to reduce their depressed feelings.

However, psychologist Michael Ternan, of the New York State Psychiatric Institute, found in a review of 300 cases of Seasonal Affective Disorder (SAD) that barely more than half of the patients treated with light in the morning recovered fully from their depressions, and the recovery rate decreased to about one-third for those who received light therapy at midday or in the evening.

Ternan hypothesized that brighter lights might be more helpful. During the winter of 1987/88 he treated 18 SAD patients with exposures of 30 minutes each morning to lights that were twenty times brighter than normal room light. The depressed patients improved substantially, usually after only a few days of treatment. The next step is to develop computer-controlled light systems for patients' rooms that will automatically simulate the natural light of sunrise on dark mornings.

Depression, in general, apparently has daily and even seasonal rhythms. Some researchers believe that one aspect of depression may be that the patient's sleeping and waking cycle goes out of phase with the day-night cycle. The patient's circadian clock is out of kilter. Studies have shown that depression symptoms are worse in the morning than in the evening and more common in the late spring and autumn.

Thomas Wehr, a psychiatrist at the National Institutes of Health, has attempted to treat depression cases by putting the patients on various sleeping schedules. He has found that the timing of the sleep is more important than the amount of sleep the patient gets. This suggests that depression can be treated by bringing the patient's internal circadian rhythms into better harmony with the day-night cycle of the real world.

One final bit of Shakespearean poetry.

When testosterone-rich Romeo swears his love "by yonder blessed moon," Juliet (who evinces a high level of estrogen) replies:

> O, swear not by the moon, the inconstant moon,
> That monthly changes in her circled orb. . . .

Ask any police officer and you will be told that people behave peculiarly when the Moon is full. Legends of vampires and werewolves hinge on the full Moon. While debunkers display statistics to show that insane behavior does not depend on the phase of the Moon, news reporters, hospital personnel, police officers, and many other people whose professions put them in contact with weird behavior insist with equal firmness that, statistics or no, people go screwy when the Moon is full.

We call crazy people *lunatics* and say they are *loony,* words derived from the Latin word for the Moon, *luna.*

Is the pineal gland sensitive to moonlight? All the hard evidence points against it. The brightest full Moon bathes the Earth with only 0.3 lux—some 400,000 times less than ordinary summer sunlight and more than 300 times dimmer than everyday indoor lighting. If the pineal gland is not activated by artificial lighting, how could it possibly be triggered by moonlight?

And yet . . .

Women's menstrual cycles are monthly, and *monthly* means, literally, "of the moon." The Moon revolves around the Earth ev-

Does the Moon actually influence human behavior? Old legends of vampires and werewolves are matched by modern claims that "lunatics" are most active when the Moon is full.

ery 27 days, 7 hours, 43 minutes, and 11.5 seconds. But because the Earth is moving too, it takes 29 days, 12 hours, 44 minutes, and 2.8 seconds for the Moon to go through all its phases (from full Moon to the next full Moon, for example). Human menstrual cycles generally average 28 days. Remember, nothing in our bodies happens without some part of the brain regulating it. Menstruation is no more automatic than picking up a pencil and writing one's name; it seems automatic because the brain function that controls menstruation is not under conscious control.

Was the menstrual cycle set in our far-distant past by the inconstant Moon? Was our third eye sensitive to moonlight ages ago? Or is this merely an astronomical coincidence? There is no way to tell, not until generations of humans live on worlds where the moons move in different cycles, or there are no moons at all. Even then our long history beneath the single Moon of Earth may have fixed this bodily cycle so firmly that it will not change.

Light has so many effects on our bodies that researchers at the National Institute of Mental Health are investigating the possibility that light therapy may help in cases of schizophrenia.

They have found that the level of dopamine in rats' brains varies seasonally, and thus must depend on the amount of light the rats' brains receive. In humans, dopamine serves several functions in the brain, including the alleviation of pain. Schizophrenia is believed to be associated with unusually increased dopamine production.

The researchers studied schizophrenic patients who had been

taken off all medication. They found that the rate of the patients' eye blinking varied with the seasons: 36 blinks per minute in spring-summer, compared to 25 in autumn-winter. Eye blinking is associated with dopamine levels in the brain. They also found that some patients appear to have increased sensitivity in the blue-pigment cones of their retinas, although this finding is not certain as yet.

If the researchers are on the right track, it may someday be possible to treat schizophrenia with light therapy, using light to control the level of dopamine in the brain and thus alleviate the symptoms of this behavioral disorder.

Researchers from many institutions are investigating the effect of light on aging. They have found, for example, that indoor lighting apparently ages laboratory mice faster than natural lighting. They believe that natural sunlight, particularly the gradual transitions between darkness and daylight that come each dawn and dusk, provides light in wavelengths that indoor lighting does not duplicate.

On the other hand, constant lighting seems to increase the longevity of hamsters. Experiments have shown that hamsters living under constant laboratory light live longer and are healthier than similar hamsters living under 12 hours of light and 12 hours of darkness.

Light—its presence or absence—affects us every minute of our lives. The way we grow and mature, our health, our sleeping, feeding and mating habits, even our moods and the ultimate length of our lives, are affected by the amount of light we receive.

We are truly creatures of the light. But what is light? What is this stuff that our eyes are sensitive to and our brains use for information?

That question has puzzled and fascinated thinking people for thousands of years. It has been a tantalizing enigma, a goal for the best minds the world has ever known, a sort of Holy Grail for scientists to pursue since the earliest days of civilization.

The quest to understand the nature of light has led curious human beings down into the innermost secrets of the atom and out to the farthest reaches of the starry universe.

II.
TO
LEARN

[Scientists] seek to investigate the true constitution of the universe—the most important and most admirable problem that there is.

—*Galileo Galilei*

9

Faster Than a Speeding . . .

WHAT is light?

It's all around us. We use it, it affects our health and our life cycles, it grows our food, we produce it artificially, sometimes we even hide from it. But what *is* it?

Long before it was possible to even begin learning the answers to that profound question, people found that pieces of glass could focus light, magnify objects, and even make the pretty band of rainbow colors that scientists now call a *spectrum.*

While the earliest "eyeglasses" were made of gemstones—the emperor Nero had a pair—hypermetropic people began using glass lenses in the Middle Ages to help them focus their near-distance vision so that they could read.

One of the greatest thinkers of ancient times used light, according to some accounts, as a weapon of war some 2,200 years ago.

In ancient Syracuse, a Greek city on the island of Sicily, Archimedes allegedly constructed giant "burning glasses" to set the ships of an invading Roman fleet on fire. Born in 287 B.C., Archimedes was truly a giant intellect. He worked out many of the principles of mechanics and said that if he had a lever long enough and a place to stand, he could move the world. It was Archimedes who discovered the principle of buoyancy when in his

In 212 B.C. Archimedes allegedly constructed giant "burning glasses" to set invading ships afire in defense of his city of Syracuse, in Sicily. When the invaders conquered the city, a soldier killed the philosopher, despite orders to spare him.

bathtub and got so excited that he ran naked through the streets of Syracuse shouting *Eureka!* ("I've found it!"). He invented a device for lifting water that is still called the Archimedes Screw, and he built many devices for the study of astronomy and other purposes.

Including war.

In 212 B.C. the expanding Roman republic invaded Syracuse. Archimedes is said to have built huge focusing mirrors that directed intense sunlight onto the Roman ships in the harbor, setting them ablaze. Many historians of science doubt this tale, but there is no doubt about the end of the story. Archimedes was slain by a Roman soldier when the city fell, even though the victorious general had given orders that Archimedes and his household should be spared the general slaughter that followed the Roman victory. But Archimedes was not in his home. A soldier came across him as he was absently drawing geometric figures in the sand of the beach and stabbed him to death.

Brilliant though he was, Archimedes had no way of seriously investigating the fundamental nature of light. He had neither the experimental tools nor the mathematical tools to get at the question. Those would not be developed for nearly two millennia. Long before Archimedes' time, and for many centuries afterward, scholars and philosophers could do no more than invent theories about light and argue over them.

Light is strange stuff. It is all around us, yet we cannot grasp it, hold it in our hands, take it apart and examine it. Under some conditions, we cannot even *see* light! Try this: Turn on a slide projector in a darkened room where the air is free of dust. Stand to one side of the projector's beam, about halfway between the projector and screen. Can you see the beam of light coming from the projector? No. But blow a puff of smoke into the air, or spray a mist from an aerosol can, and you will see the beam clearly delineated. Some of the light from the beam is scattered by the particles we have put into the air and reaches our eyes.

We do not see light unless it is aimed directly into our eyes. A beam of light can pass right by, but unless it reflects off something so that the light enters our eyes, we cannot see the beam. This fact, plus others, led to the conclusion that light travels in straight lines.

To the ancients, however, the question was not about the nature of light so much as the nature of vision. They recognized light as a "carrier" of vision, but it was vision itself that intrigued them. How do we see? How does vision work?

It seemed obvious that something connects between the eyes and the thing being viewed. That something must be light. But does the light come from the eyes, or from the object under scrutiny, or from someplace else?

Some thinkers actually believed that light originated in the eyes. They thought that when we look at something, beams of light come out of our eyes and scan the object we are looking at. In this view, our eyes are rather like searchlights illuminating the world around us or like radar transmitters sending out invisible signals.

The earliest known writing on the subject is by Euclid and dates back to 280 B.C. The man who is credited with inventing geometry, Euclid, following the ideas of the philosopher Plato, wrote that rays of light originate in the eye and strike the object being viewed. The speed of light must be very high, Euclid believed, because you can close your eyes (thus making the things you are looking at disappear!) and then, when you open them again, even the distant stars appear instantly.

A century later Hero of Alexandria reasoned that if light travels with *infinite* speed, then it must move in straight lines. He was half right.

You can see the flaw in this approach every time you stumble on something in the dark. If light originates in our eyes, how come it doesn't work in the dark? Scratch the light-from-the-eyes theory.

Well then, the reasonable alternative is that light originates in the objects we see. They radiate light and our eyes receive it. Better, but still not good enough. Again, how do you explain that bump in the night? If light originates with the coffee table, why doesn't the damned thing let me see it when I turn off the living room lamps?

Ahah! Light comes from certain sources, bounces off objects, and enters our eyes.

We see the objects around us by the light those objects reflect— unless the objects themselves are producing light. These words you are reading, these symbols on this page, you see by the light they reflect into your eyes. The light bulb that provides illumination (or the Sun, if you are outdoors in daylight) is self-luminous; it emits light.

Right. The original source of light was the Sun, all 114,000 lux of it blazing up there at midsummer noon. Then our ancestors tamed fire, which provided light when they needed it.

Among the first thinkers to recognize that light originates in sources that illuminate the objects we see was an Arab from the fabled city of Baghdad in what is now Iraq. While Europe groped through the Dark Age following the collapse of the Roman Empire, the new religion of Islam rose like a whirlwind out of the Arabian Desert and quickly swept westward as far as Spain and eastward through India and beyond. A great civilization arose in the Middle East, where much of the knowledge of the ancient Greeks, forgotten or ignored in Europe, was preserved and enlarged upon.

Abu Ali Mohamed ibn al-Hasan Ign al-Haytham, known in Europe as Alhazen, wrote seven books on optics around the year 1000. He recognized that light originates in sources, like the flame of a candle, reflects off objects, and then enters our eyes to create vision. He hypothesized that light does not travel at infinite speed, although its speed must be very great. He also correctly surmised that light travels more slowly in dense media, such as water, than it does in more rarefied media, such as air.

Alhazen used a delightfully simple device, the camera obscura, to prove that light travels in straight lines. *Camera obscura* literally means, in Latin, "dark room." A chamber (it could be as large as a room or as small as a box you can hold in your hand) is so tightly closed up that it is darkened completely, so that the only light allowed into it comes from a pinhole in one wall. The light from that pinhole falls on the wall opposite it, which is usually painted white, while all the other walls are black.

The image carried by the light through the pinhole will appear on the white wall upside down! Try it for yourself. No lenses are involved. Picture the geometry of it. Suppose you point the

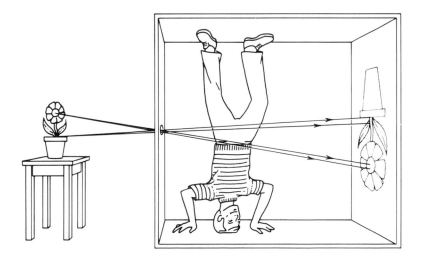

The camera obscura *was used by the Arab scientist Alhazen around* A.D. *1000 to show that light travels in straight lines. Like the human visual system, the camera obscura produces an image that is upside-down.*

pinhole of a camera obscura at a tree. The light rays coming from the top of the tree will go through the pinhole *in a straight line* and strike the whitened wall near its bottom. The rays from the bottom of the tree will hit near the top of the camera obscura's white wall.

An amusing toy, and in Europe some five centuries after Alhazen, Leonardo da Vinci and other Renaissance artists began using the camera obscura as an aid to drawing. Pose a subject and then look at it through a camera obscura (or pinhole camera, as it is often called). Place drawing paper where the image falls and you can trace the subject nicely. More sophisticated versions used a mirror to rereverse the image and make it right side up again.

A little more than four centuries after *that*, two Frenchmen, Joseph Nicephore Niepce and Louis Jacques Mande Daguerre, hit upon the idea of placing a plate containing light-sensitive chemicals at the back of a camera obscura. Photography was born, and to this day we call the instrument we use for making photographs after the old Latin word for "chamber": *camera*.

Alhazen thought of light as a stream of particles that bounce off an object—the phenomenon of *reflection*. He also noted that light rays seem to bend when they go from a medium of one density, such as air, to a medium of a different density, such as water: For example, note how a stick held halfway into a pool of water seems to be bent. This is the phenomenon of *refraction*, and Alhazen correctly attributed this to the fact that light travels at different speeds in media of different densities.

In the world of science there is an old maxim: Publish or perish. This means that if a scientist does not publish papers describing his or her research, the scientist will not gain the recognition of his or her peers. No recognition means not only that the work itself will be ignored; it also means no promotions, no funding to continue the work, and ultimately, no career in science.

This maxim holds true for ages past. There must have been many, many ancient philosophers who pondered the nature of light. But the only ones we know about are those precious few whose writings have survived the vicissitudes of time. When the great library at Alexandria burned, in Julius Caesar's time, much of the ancient world's scientific knowledge went up in smoke.

We know of the work of Alhazen and a few other Arab sci-

Light rays seem to bend when they go from one medium to another, such as from air to water. This phenomenon is called refraction. *It is the principle behind the light-bending powers of lenses.*

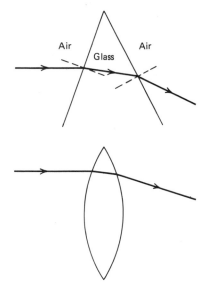

entists because they wrote down their thoughts—and their books somehow survived the ravages of time and the savagery of ignorant, prejudiced men who all too often willingly destroyed the records of the past because it did not agree with their current biases. Of all the self-inflicted wounds the human race has suffered, the most dangerous and deadly is the burning of books. It is a lobotomy that cuts off our own past and offers nothing, literally *nothing,* in return for the loss. Those who burn books, or attempt to prevent people from reading certain books, are trying to control the minds of their people. They should go to the Jefferson Memorial in Washington, D.C., and read the quotation carved into the inner circumference of its beautiful dome: "I have sworn on the altar of God eternal hostility against every form of tyranny of the mind of man."

You cannot be a Jeffersonian and a book burner. You cannot be a free human being in a society that bans books.

The major ancient thinker who stands out in the history of science is Aristotle. Why? Not because his theories were correct. Much of what Aristotle wrote about the physical sciences has been proved to be terribly wrong. But Aristotle's works survived. He wrote treatises about every aspect of human thought. His contemporaries wrote too, but most of their writings have been lost to us, deliberately destroyed by zealous bigots when Christianity was made the official religion of the Roman Empire or later casually destroyed by invading barbarian tribes in the collapse of the ancient world that we know as the Dark Age.

Aristotle endured. The others published but perished anyway. Although Aristotle was mostly wrong in his ideas about the physical sciences, he became the authority figure throughout Europe because the Church of Rome accepted his teachings. For more than a thousand years Aristotle was it. The Church that held all of Europe in its sway regarded him as the authority on natural philosophy for more than ten centuries.

Born in 384 B.C. in Macedonia, Aristotle was, among other things, tutor to the young Alexander the Great. His interests were universal, ranging from philosophy to logic to politics to natural philosophy (what we today call science). He was a brilliant thinker, and although much of his scientific work has been shown to be flawed, his influence on Western thought has been enduring.

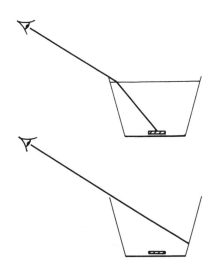

Light is refracted when it goes from air to water. If the cup is filled with water (top), the coin can be seen; if the cup is filled with air, the coin cannot be seen by an observer at the same angle.

Aristotle never distinguished between light and color. To him, the problem of vision was a problem of perceiving colors. "Whatever is visible is color," he wrote, "and color is what lies upon what is in its own nature visible." Moreover, Aristotle believed that all colors are formed from mixtures of white and black.

Like most of Greek natural philosophy, Aristotle's ideas were flawed by two basic misperceptions. First, the Greeks tended to be idealists. That is, they set up a philosophical ideal of what the world should be (as they saw it) and then tried to explain the phenomena of the real world in terms of the ideal.

For instance, they firmly believed that everything in the heavens was perfect, unlike our highly imperfect world here below. Thus, the planets must move in circular orbits, because the circle is a "perfect" geometric figure. This basic idea held back the progress of astronomy for more than 2,000 years! It was not until the early seventeenth century that Johannes Kepler proved that planets travel in lopsided elliptical orbits. That discovery of "imperfection" in the heavens paved the way for swift and enormous scientific progress on Earth.

The second flaw that hounded Greek thinking was their lack of experimental verification of their ideas. For the most part, Greek philosophers expounded their theories and then argued over them. They did not try experiments to test whether or not their theories were validated by the actual behavior of the real world.

Even the brilliant Archimedes regarded the mechanical devices he invented as somewhat beneath his dignity. Ideas were the important thing to the Greeks, perhaps overly influenced by the concepts of Plato, who taught that ideas are more "real" than physical objects. Getting your hands dirty was work for slaves, not gentlemen-philosophers.

For example, consider the matter of falling bodies. Most Greek philosophers taught that a heavy object will fall to the ground faster than a light object. In fact, when you think about it, it seems obvious. Yet it is not so.

A feisty Italian from Pisa, Galileo Galilei, got the notion that heavy objects and light objects fall to the ground with the same speed. He could have spent his life arguing with fellow university sages about this seemingly absurd notion. But instead he carried two metal spheres of different weights up to the top of a tower and dropped them. They hit the ground at the same time. In a few seconds, Galileo proved that he was right and the whole body of argument that had been made for thousands of years was wrong.

Legend has it that Galileo used the famous Leaning Tower of Pisa for his experiment. This is undoubtedly wrong, since the work was done around 1604, and he had been a professor at the University of Padua since 1592.

Experiment. Test. Trial. What Galileo was doing was helping to create the most powerful force ever developed by the human mind: an idea, a set of techniques and attitudes, that we now call *the scientific method*. The key to the scientific method is this: All ideas must be tested by experiment or by observation of the natural world. Otherwise you don't have science, you have talk.

The great English physicist of the nineteenth century Michael Faraday once defined science as "to make experiments and to publish them."

By testing, by experiments deliberately designed to confirm new ideas or break them, the scientific method has amassed an incredible treasure of knowledge for the human race in a mere few centuries. Our modern world has been created by science and its offspring technologies.

Galileo was born in 1564, the same year that Shakespeare was born. A great year for both art and science. In fact, the flowering of arts and learning that we call the Renaissance had by that time brought much of Europe back to the standard of living that the people of ancient Rome had enjoyed. Average life expectancy had increased to about 33 years.

In the slightly more than four centuries since then, scientific knowledge has allowed us to transform the world. Slavery has been replaced by machinery. Life expectancy has risen to nearly 75 years in the industrialized (i.e., scientifically advanced) nations. Diseases that cut us down in infancy or young adulthood have been eradicated. Democracy and individual liberty rest on the base of scientific achievements that have freed the human mind and spirit from the dogmatism and tyrannies of the past. The scientific method has served us well.

Like many thinkers before him, Galileo was curious about the nature of light. Did it truly travel at infinite speed? He tested the possibility with an ingenious experiment that was, unfortunately, doomed to fail.

He put two men on hilltops some distance from each other. He measured the distance between them carefully. Each man had a lantern that was shielded so that its light could not be seen unless the shield was dropped. One of the men (undoubtedly Galileo himself) had a clock.

The man with the clock was to drop the shielding from his lantern and start the clock at the same instant. The other man, on the distant hilltop, was to drop the shielding on his lantern as soon as he saw the light from the first lantern. When the first man saw the light of the other lantern, he was to stop his clock. Galileo intended to repeat this experiment at constantly larger distances until he found a reliable value for the speed of light.

A clever experiment, but it failed. The time between opening the first lantern and seeing the light from the second one was so brief that it could not be measured, especially with the crude clocks available in the early seventeenth century. No matter what the distance, light was too fast to measure.

Galileo might have concluded that the speed of light was infinite. But, good scientist that he was, he realized he could not really say that. All he could say was that the speed of light is so great that he was unable to measure it.

The speed of light was not finally measured until 1676. Interestingly, its measurement was made possible by one of Galileo's greatest discoveries.

It happened this way. Hans Lippershey, a Dutch maker of eyeglasses (then called spectacles), hit upon a marvelous invention: the telescope. By placing glass lenses of the proper type in a tube, he could magnify distant objects and make them appear as if they were nearby. On October 2, 1608, Lippershey formally offered his invention, which he called a *kijker* ("looker"), to the government of Holland for use in war. Just as in today's society, the military has long been among the first customers for any new technology.

Eyeglasses, or spectacles, had been in use in both Europe and China since at least the thirteenth century. Historians argue over whether they were invented in China and carried to the West, or vice versa. Certainly the travels of Marco Polo and other traders played a role in the exchange. There was no science of optics as yet. Spectacle makers learned by trial and error that lenses of certain shapes helped people with failing eyesight to see better.

(It was not until 1784 that Benjamin Franklin invented bifocals, in which two different lenses, one for close vision and one for far, were fastened together in the same frame.)

Galileo heard about Lippershey's invention some months after the fact. Knowing nothing except that it involved a pair of lenses set into a tube, he built his own telescope. Galileo quickly deduced what kind of lenses were needed, probably ground his own, and (if legend can be believed) sawed a chunk from a church pipe organ to serve as the tube.

Then Galileo displayed the difference between talent and gen-

A page from the notes Galileo made of his first telescopic observations of the planet Jupiter and its four largest moons, known to this day as Jupiter's Galilean satellites.

ius. Instead of using his telescope to snoop on his neighbors or spy on Padua's enemies, he turned it to the night sky. He did not invent the telescope, but he did invent modern astronomy.

Among the many discoveries Galileo made that eventually shattered the old medieval views of the world (and got him imprisoned by the Church) were the moons of Jupiter. In 1609 he discovered that the planet Jupiter was accompanied by four moons. This was the first time any object in the heavens was unequivocally shown to revolve around a body other than the Earth. Sixty-six years earlier, Nicholas Copernicus had theorized that the Earth was not the center of the universe but orbited around the Sun. The Copernican theory was not widely accepted even by scholars at that time and was forcibly resisted by the Church.

The four moons of Jupiter that Galileo discovered are still called the Galilean satellites, although modern observations— including those made by spacecraft such as *Voyager*—have shown that Jupiter has at least sixteen moons, as well as a faint ring of pulverized particles, circling it. (See Plate 8 following page 78.)

Sixty-seven years after Galileo's first telescopic observations, the Danish astronomer Ole Roemer actually measured the speed of light by using observations of Jupiter's satellites.

By Roemer's time the orbits of Jupiter's Galilean satellites had been calculated to such a precision that astronomers could predict with good accuracy when each of the four moons would pass behind Jupiter's flattened disc and be eclipsed by the giant planet. To Roemer, this was a "lantern" that could be used to measure the speed of light.

By timing the eclipses of Jupiter's moons from position (1) and position (2), Ole Roemer measured the length of time it took light to travel the diameter of Earth's orbit and thus calculated the speed of light to a rough accuracy.

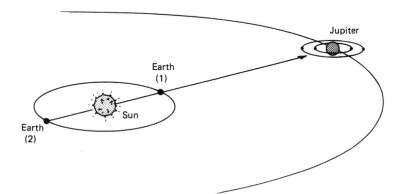

He found that when the Earth was in the part of its orbit that brought it closest to Jupiter, the eclipses took place 11 minutes earlier than predicted. When Earth and Jupiter were farthest apart, the eclipses were 11 minutes later than the predictions. Roemer concluded that the 22-minute difference represented the time it took light to cross the diameter of the Earth's orbit around the Sun. The early eclipses were due to the fact that the Earth was close enough to Jupiter to receive the light from the planet and its moons sooner than the calculations predicted. The late eclipses happened because the light had to travel an extra distance since the Earth was farther away.

To calculate the speed of light, Roemer needed only to know the diameter of the Earth's orbit around the Sun and divide by 22 minutes. However, in the seventeenth century, the diameter of the Earth's orbit was not known to any reasonable accuracy. Astronomers had been able to lay out the geometry of planetary orbits very nicely, but they did not have a truly accurate yardstick to measure any of the interplanetary distances very well. Moreover, Roemer's clock was not as accurate as it needed to be for a precise measurement of light's speed; we know today that light crosses Earth's orbit in 16 minutes, 36 seconds, not 22 minutes.

As it was, Roemer got the technique right, but the numbers he came up with were not so right. He arrived at an estimate of 132,000 miles per second. Astoundingly fast by any earthly value. So shockingly fast, in fact, that many of his contemporaries refused to believe it could be possible. Nothing could move that fast, they insisted. But Roemer's figure was not fast enough.

With modern tools the speed of light has been measured at 186,282.3959 miles per second. Call it 186,000 miles per second, or 300,000 kilometers per second.

Fast! The light you see when you look at the Moon left the lunar surface about 1.3 seconds before it reached your eyes. The Sun is eight light-minutes away; it takes light eight minutes to cross the 93-million-some miles between Sun and Earth.

The stars are light-*years* away. As we saw in Chapter 2, a light-year is a measure of distance, not time, and amounts to roughly 6 trillion miles.

As a dear friend of mine is wont to say, "The mind bagels!"

Astronomical distances are, well, astronomical. It's virtually impossible for the mind to grasp such enormities. Here is one method that I have found to help to visualize these vast distances.

The distance between the Earth and the Sun averages roughly 93 million miles. Draw a map, in your imagination, in which that huge distance is shrunk down to one inch. Now, on the scale of one inch equaling 93 million miles, how far away must we place the nearest star, Alpha Centauri? The answer is, 4.3 miles. And that is the *closest* star to our Solar System. The distances inside the Solar System can be pictured as inches, the distances among the stars as miles.

Galileo spent his final years, feeble and going blind, under house arrest by the Inquisition for his heretical views. The year that he died, 1642, Isaac Newton was born in England. (Mystics, take note.)

Of all the scientists who have pursued the study of nature, Newton ranks at the very top both in the regard of his fellow scientists and in the eyes of historians, equaled perhaps only by Albert Einstein.

Newton's contributions to our understanding of the universe are so fundamental, in so many fields, that it is easy to see how he gained such esteem. He formulated the basic laws of motion, of gravity, and of optics. He invented calculus,* an indispensable calculational tool although the bane of many a student.

More than any other contribution, though, Newton brought to science the idea of *universality*. In particular, his concept of gravity as the force that affects both the fall of an apple and the orbit of the Moon around the Earth was the first time that human thought had found a link between what happens here and what happens elsewhere in the universe. It was Newton, more than anyone else, who convinced his fellow scientists that the physical laws they uncovered were universal laws that work the same on Earth as they do on the most distant stars.

This was an extremely important step in the history of science. It led Einstein to say, three centuries later, "The eternal mystery of the world is its comprehensibility." In other words, we

---

*Calculus was also invented, independently and at about the same time, by Gottfried Wilhelm Leibniz. Newton's admirers (and the great Sir Isaac himself) bitterly accused Leibniz of plagiarism. However, it often happens that discoveries or inventions are made nearly simultaneously by people working in the same area.

can understand the way the universe operates. There is no separation between the "perfect" heavens and the "imperfect" realm of humankind. The universe is one. We are as much a part of the cosmos as a star is. Indeed, as we shall see later on, we are made of stardust.

They say that every cloud has a silver lining. In Newton's case, the cloud was the Great Plague that swept England in 1664–65, killing 70,000 in London alone. In 1665, shortly after Newton received his bachelor of arts degree from Cambridge University's Trinity College and had begun teaching there, the school closed its doors for nearly two years. The plague was far worse in the crowded, unsanitary cities than in the countryside. At the age of 23 Newton was forcibly retired to his mother's estate in Lincolnshire.

It was there, while sitting under an apple tree on a summer afternoon, that he got the idea of universal gravitation. During the months that the university was closed Newton worked out the principles of gravitation and motion that would become the subject matter for his great book, the *Principia.** He also produced the first truly scientific work about light, which became the subject of his *Opticks.*****

(Some two-and-a-half centuries later, Abraham Flexner would create the Institute for Advanced Study at Princeton as a place where brilliant scholars could work creatively without the need to teach classes or do administrative work. He was influenced, no doubt, by the example of Newton's enforced "idleness.")

Newton did not merely theorize about light. He experimented. Turning one of the rooms of his mother's house into a virtual camera obscura, he allowed light to enter through a pinhole in one of the window blinds. He aimed this slim shaft of light at a glass prism. The light went through the triangular piece of glass and spread out into the lovely rainbow pattern of the spectrum on its other side. There was nothing new in this. People had been doing it for ages. And for ages they had argued over whether natural sunlight actually contained all those beautiful colors

*The *Principia,* not published until 1687, was written in Latin. Its full title is *Philosophiae Naturalis Principia Mathematica,* or *Mathematical Principles of Natural Philosophy.*
**Opticks,* written in English and spelled in the manner of the times, was not formally published until 1704.

bundled up together or whether the glass somehow "stained" the light, colored it.

Newton went a critical step further. To determine whether the colors were inherent in sunlight or were somehow caused by the glass, he let one of the pure colors coming out of the prism—yellow, as it happened—pass through a second prism.

Nothing but yellow light came out of the second prism. This proved conclusively that the colors were not caused by the glass. Natural sunlight—which Newton called white light—actually contains all the colors of the rainbow.

He found another proof of this by allowing the spread-out spectrum from one prism to pass through another prism oriented

Sir Isaac Newton's experiments led to the first scientific understanding of the nature of light.

in the opposite way from the first. The rainbow merged into a shaft of white light on the other side of the second prism.

Newton's crucial experiment dealt with the phenomenon of *refraction,* the bending of light as it travels through media of differing densities.

The natural phenomenon of the rainbow, one of the most beautiful sights our eyes can behold, singled out in Genesis as a reminder of God's mercy—the rainbow can now be seen as a natural example of refraction. As sunlight passes through raindrops the light is bent, refracted, by the water and spread by these natural prisms into the glorious colors that we see.

Newton correctly interpreted refraction as the result of the different colors of light being bent by different amounts as they passed through the glass. In air, all the colors of white light travel at the same speed and remain together. In the glass of the prism, however, the colors are spread apart because they are bent—refracted—at slightly different angles. (See Plate 9 following page 78.)

He also stated quite clearly that *color* is actually a matter of perception. While it is possible to examine and experiment with the physical properties of light, what we perceive as color depends on what goes on inside our heads. You and I may agree that a certain light is red, but we will never be able to know if we truly see exactly the same color. Try describing red to a person who has been blind from birth. It cannot be done. Or try describing the color of something to someone over the telephone. Listen to what you say. You will find that you offer *comparisons* of the color: "It's not quite as dark as navy blue, more like a royal blue, almost, but with more of a reddish tinge to it."

We describe colors in terms of other colors—or in abstract scientific terms that describe the physical nature of the light but do not describe the color at all.

Newton wrote, "Indeed [light] rays, properly expressed, are not colored. There is nothing else in them but a certain power . . . to produce in us the sensation of this or that color."

About a century later, in 1800, the astronomer William Herschel discovered that there is more to the spectrum of sunlight than the rainbow of visible colors. Trying to measure the energy contained in sunlight at different colors, Herschel passed a thermometer slowly across the spectrum coming from a glass prism. To his surprise, he found that the highest temperature reading came from the region beyond the red, where no color at all was visible.

There was obviously some invisible form of radiation coming from the Sun that was carrying most of its heat energy. We call that radiation *infrared;* it is the radiant heat that we cannot see but our skin can feel. As the nineteenth century unfolded, other scientists found other forms of "light" that are invisible to our eyes: ultraviolet, radio waves, X-rays, and gamma rays.

Newton, of course, dealt only with visible light, the stuff our eyes can detect. Like Galileo, who realized that he could not measure the speed of light, Newton realized that he could not definitively determine the basic nature of light. However, his experiments showed that light behaved as if it were composed of a stream of tiny particles. He knew of Roemer's calculation of the speed of light; even though it was not precisely correct, it was good enough to show that light traveled extremely fast.

To support the idea that light consists of a stream of submicroscopic particles, Newton pointed to the observable fact that light travels in straight lines. You can demonstrate this for yourself. Shine a bright light against a light-colored wall or screen. Place an object between the light source and the wall. Now take strings and run them from the light source to the edges of the object's shadow on the wall. You will find that the strings form absolutely straight lines that go from the light source to the edges of the object itself and on to the edges of the shadow.

Every time.

Well, almost . . .

10

Waves of Light

ALTHOUGH NEWTON could be fiercely possessive about his work, to the public he usually tried to present a becomingly modest appearance. When praised about his towering achievements, he replied, "If I have seen farther than other men, it is because I stood on the shoulders of giants." He also wrote:

> I do not know what I may appear to the world; but to myself I seem to have been only like a boy playing on the seashore, and diverting myself in now and then finding a smoother pebble or a prettier shell than ordinary, whilst the great ocean of truth lay all undiscovered before me.

An interesting metaphor. When little children play on the seashore they often throw the pebbles or shells they find into the water. And when a pebble hits the water, it creates little waves that spread out in a circle all around it.

There were scientists who, even before Newton, believed that light is *not* a flow of tiny particles but a stream of waves, like the waves of surf rolling in on a beach or the waves of sound that our ears detect.

In 1672 Newton was elected to the Royal Society, one of the oldest and most prestigious scientific societies in the world. When he reported on his work in optics (*Opticks* would not be published for another 32 years), he was immediately challenged by Robert Hooke.

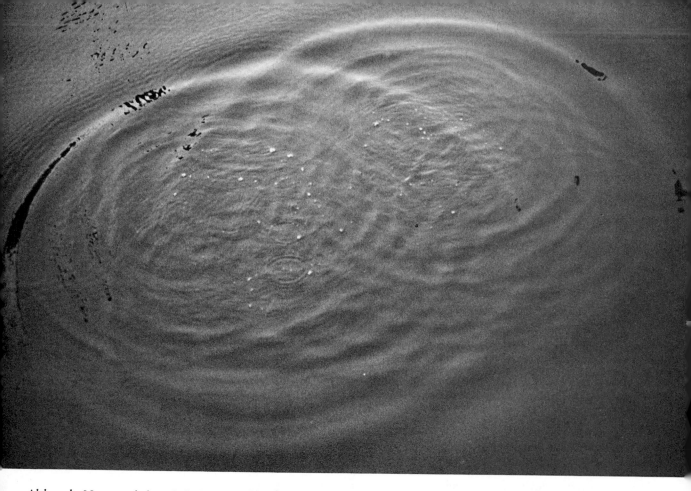

Although Newton believed light was probably a stream of ultrasmall particles, other scientists insisted that light consisted of waves, somewhat like waves in water or waves of sound.

Hooke passionately believed that light was a wave phenomenon, like sound waves. One problem with that idea is that waves need a medium in which to propagate. Water waves travel through water; sound waves travel through all sorts of media, including air. Since it was obvious that light traveled through interplanetary space, Hooke believed that space was not empty but was filled with an invisible substance he called ether. In later generations, the "luminiferous ether" became a firm part of established physical theories. Belief in the ether had world-shaking consequences—even though it never existed in reality.

Belief is one thing. Proof is another. Hooke was not the careful experimenter that Newton was, nor was Hooke as deep a thinker (who was?). Yet he attacked Newton so vehemently that a lifelong enmity developed between the two of them.

The Dutch astronomer Christiaan Huygens took up Hooke's ideas and developed a mathematical treatment that showed how light waves are bent by a prism or a lens. Huygens was a many-faceted man: He developed an improved method for grinding lenses; he observed that the planet Saturn is surrounded by bright rings and discovered Saturn's largest moon, Titan; he improved on Galileo's work with pendulums and built the first grandfather's clock, which ushered in the age of accurate time-keeping.

Huygens believed that light was a wave phenomenon, like the ripples on a pond when a pebble is dropped into the water. But whereas the water waves on a pond's surface are two-dimensional, the waves of light coming from a candle (for example) are three-dimensional: They expand outward spherically, like swelling balloons. His mathematics, based on the wave theory, explained refraction at least as well as Newton's particle concept.

Newton himself realized that there were certain things about the behavior of light that could not be easily explained by the particle theory. Indeed, in *Opticks* he carefully stated that the concept of light being composed of submicroscopic particles was not proved; it was merely an attempt at an explanation based on the available evidence.

While Huygens was a giant among the era's scientists, Newton was a supergiant. His reputation had become unassailable, and—as happens so often—his supporters took a much stronger stand against the wave theory than Newton himself did.

It must be added here that the latter part of Newton's life was marked by a turn toward mysticism, and some modern investigators believe that he may have suffered brain damage from heavy metal poisoning. Newton spent years working with large amounts of mercury, in part for his studies of optics, and it is possible that he breathed in enough mercury vapor to eventually kill him.

Newton's followers backed the particle theory so strongly that it seemed unassailable. This is not to say that everyone accepted Newton's work without criticism. Even after both Hooke and Newton died, there was discussion and even argument. But gradually the particle theory, with Newton's enormous reputation attached to it, became *the* established theory.

Still, there were doubts. In 1810 the German writer and philosopher Johann Wolfgang von Goethe published his *Theory of Color*, in which he attempted to prove Newton wrong. He failed, principally because he reverted to philosophical analogies instead of conducting experiments that could provide data that contradicted Newton's findings. The author of *Faust*, a man who dramatized the concept that the thirst for knowledge is a path to damnation, tried to undo experimental evidence with words. The evidence prevailed.

However, Goethe's words about how human beings perceive colors did serve as a forerunner of the modern science of perceptual psychology.

In science, though, any concept can be overthrown. No matter how well established an idea may be, no matter how prestigious the people who support it, it can be overthrown. Science is very different from other areas of human endeavor in this regard; the establishment can be blown away on the strength of a new piece of information. There is no permanent priesthood in science. Science is an ongoing process, a constant search for truth.

This is what makes many people, and particularly politicians, so uneasy about science and scientists. Nothing is certain forever. Nothing is set so firmly in concrete that it cannot be uprooted and supplanted. Most people find that unsettling. Most politicians find it unnerving. They want eternal verities. Hitler provided eternal verities. So did the Spanish Inquisition and Josef Stalin's regime.

Despite the fact that the authority of Sir Isaac Newton's reputation was protecting the particle theory of light, that theory was eventually supplanted. One of the major actors in the overthrow was a failed physician. There was a chink in the armor of the particle theory, and it had actually been noticed by a relatively obscure Italian physicist, Francesco Maria Grimaldi, at just about the time that Newton was first presenting his work to the Royal Society and stirring the wrath of Hooke.

Grimaldi noticed that light does not always travel in strictly straight lines. Under certain conditions, light rays bend slightly around an obstacle, a phenomenon called *diffraction*.

For more than a hundred years this uncomfortable little fact lay ignored. It was uncomfortable because it could not be explained by the particle theory. If light is composed of particles, and if those particles travel in straight lines, how come they can bend around corners? The phenomenon of diffraction could be ignored because the evidence for it was very slight; the bending that Grimaldi had noticed was quite small, and his work was not widely read.

Then came Thomas Young, an English physician who was more interested in research than treating patients. A child prodigy who could read by the age of two and had read the entire Bible twice before he was six, Young discovered in 1801 that astigmatism is caused by a roughening of the surface of the cornea and, more than a decade later, helped to translate the Rosetta Stone, which opened the way to understanding the writings of ancient Middle Eastern civilizations.

In the meantime, he demolished the particle theory of light.

Young's interest in the human eye led him to an interest in light itself. In 1803 he performed a classic experiment. He passed a beam of light through two closely spaced pinholes and then allowed the light to fall on a screen. Instead of two spots of light, the screen showed bands of alternating light and darkness. These bands are called *interference patterns*. There is no way that streams of particles could produce such patterns. Only waves can. Young made the comparison between sound waves produced by tuning forks and waves of light.

You can duplicate Young's experiment for yourself. Get a flashlight and a pin. Then take three pieces of white cardboard, the sort of board that laundries place in men's shirts.

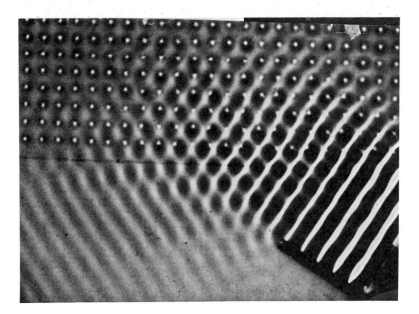

Interference patterns cannot be explained by the particle theory of light. Only waves can produce such patterns of alternating light and dark.

Place the flashlight close enough to the first cardboard so that it blocks the entire beam of light. Set up the second cardboard about a yard away. Now carefully put a pinhole into the first cardboard. Turn off all the room lights, leaving only the flashlight on.

The second cardboard will be illuminated by the tiny shaft of light coming through the pinhole. It will be dimly lit, but the light will be evenly distributed across the cardboard. So far, the particle theory explains what can be seen. Streams of particles coming through the pinhole move outward in straight lines to strike the second cardboard.

Now put two pinholes in the third cardboard, close together enough so that the light beam shines on them both simultaneously. Place the third cardboard between the first two. The light now travels from the flashlight, through the first cardboard with its single pinhole, then through the next cardboard with the two pinholes, and finally falls on the rearmost cardboard.

What do you see on that rear board? The light no longer falls evenly across it. Instead, it creates a checkerboard pattern of bright

Thomas Young's experiment showed that light filtered through pinholes in a screen creates interference patterns like ripples in a pool of water.

areas and dark areas. You have created an interference pattern, just as Young did in 1803.

You can see interference patterns on a sunny day at the beach. Go into the water up to your ankles and watch the interplay of light against the sand when a wave runs back toward the sea. The ripples of water create patterns of light and dark against the sand.

Young's experiments and his conclusion that light is a wave phenomenon did not sit well with those who backed the particle theory. They dismissed Young's work as unscientific and even "un-English." Yet about a decade later a French physicist, Augustin Jean Fresnel, not only repeated Young's work but extended it so solidly that the particle theory ultimately collapsed.

It collapsed because the wave theory *worked.* The wave theory explained observable phenomena that the particle theory could not explain, in addition to explaining all the phenomena that the particle theory did explain. Science is ultimately a democratic endeavor, where the "votes" are factual observations of nature. If your "candidate" theory explains more of nature than another candidate, then you win the election. But there is always the chance of a new election, based on new observations of nature or reinterpretations of existing observations.

The wave theory outvoted the particle theory, thanks mainly to Young and Fresnel. (But there was to be another "election" in the twentieth century.)

Waves have certain physical properties to them whether they are waves of sound, water, light or something else. Waves have *crests* (their high points) and *troughs* (their low points). They also have:

1. Wavelength: the distance between crests (or between troughs, if you're a pessimist. Either measurement is fine, as long as you are consistent and don't mix crests with troughs).

2. Frequency: The number of waves that pass a given point within a fixed period of time.

3. Amplitude: This one is a little tricky. Amplitude is measured as half the distance between crest and trough. If you imagine a line of waves washing up on a beach, their amplitude would be the distance between their crests and the level of the undisturbed water.

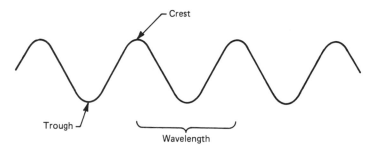

All waves have crests, troughs, wavelength, amplitude, and frequency.

Light travels very fast: slightly more than 186,000 miles per second in vacuum. Nothing in the universe is faster. In glass, light travels "only" two-thirds as fast, or nearly 124,000 miles per second.

Wavelengths of light, as we have seen, are extremely small. Visible light ranges from about 400 to 700 nanometers.

Frequency is expressed in the unit called *hertz,* which is equivalent to one cycle per second, or one wave passing the measuring point each second. The current used in lighting and other electrical equipment in the United States oscillates at 60 cycles per second, or 60 hertz (abbreviated Hz). Humans can hear sound waves, which propagate through air, from about 16 Hz to 20,000 Hz, or 20 kilohertz (abbreviated kHz). Light waves, as you might expect, have enormously higher frequencies, on the order of 100 million million (10^{14}) Hz.

The wave theory of light made sense of colors. It was already known that when white light goes through a prism it comes out spread into a spectrum of colors. Now it became clear that each different color represents a slightly different wavelength of light, and each wavelength is bent (refracted) at a slightly different angle in the prism. Red light, for example, lies in the wavelengths from roughly 650 to 700 nanometers. Blue light is from 450 to 500 nm.

Leaving aside the particle versus wave argument for a moment, let us take a look at some of Newton's painstaking experiments dealing with color.

We will be dealing here with pure colors and their mixing. By a pure color, I mean a ray of light of one single color, as it comes from a prism. Newton called such a ray homogeneous light. Physicists today call it *monochromatic* light; that is, light of

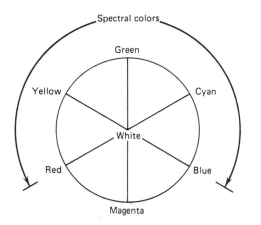

A modern version of Newton's color wheel. Red, blue, and green are primary colors; the colors opposite them on the wheel are their complementaries.

one color only. Instead of the broad mixture of wavelengths from 400 to 700 nm that make up white light, these experiments dealt with narrower wavelengths, the 50 to 100 nm spreads between one individual color and another.

Newton found that when he mixed two beams of monochromatic light of different colors, he got a third color. Mixing pure red and pure green produced yellow, for example. If he passed the yellow light thus produced through a prism, it broke up into red and green again.

We know now that the color receptors in our retinas are sensitive to red, blue, and green. Little wonder that Newton and the other scientists who studied color phenomena found that there are three "primary" colors—red, blue, and green—and all the other colors of the rainbow could be obtained by mixtures of these three.*

The science of color mixing began. Artists, of course, had been mixing pigments since the Stone Age. They worked with what they could find, colored clays and bits of crumbling earth, bright juices squeezed from berries. By the time of the Renaissance, the painter's palette had become sophisticated enough to yield the glowing colors of Titian and Raphael and the somber hues of Rembrandt.

---

*Newton dealt with light. As we shall see in Chapter 12, things are different when we deal with pigments.

Now scientists began unraveling the reasons behind color mixing. And technicians began adapting their knowledge for practical use. Color printing arose and was improved to the point where today even cheap newspapers can be printed in a rainbow of hues. Color motion pictures and television broadcasts, even colorization of films that were originally shot in black-and-white, are now commonplace.

While Newton pioneered the scientific understanding of color mixing, the wave theory of light gave his empirical studies a firm theoretical underpinning.

A word about that word *theory*.

In everyday usage, *theory* has come to mean about the same as *hypothesis:* that is, an informed guess, an unproved set of assumptions. In science, *theory* means something very different.

To a scientist, a theory is an organized set of ideas, a structure that explains a wide range of observations—and also predicts new observations. If you think of an individual observation of nature as one of Newton's seaside pebbles, a theory is a detailed map that shows where the pebbles were found, what their relationship is to the ground around them, and where new pebbles can be located.

For example, Charles Darwin gathered a vast array of observations about plants and animals—observations that ranged from the songbirds of the Galápagos Islands to the facial muscles that allow us to smile. Out of that multitude of observations he derived the idea of evolution based on natural selection: Organisms change in response to the environments in which they exist. This is called the theory of evolution. It is one of the greatest products of human thought. It made sense out of the entire study of biology and allowed scientists to understand how their different observations are related to one another. Moreover, it allowed scientists to predict where and how they could make new observations, learn new facts about nature.

Yet, because of that word *theory,* many people believe that scientists themselves are not certain of Darwin's concept. "Evolution is only a theory," they say, meaning that it is an unproved hypothesis. Not so. Darwin's concept of evolution is the central guiding path for the entire science of biology and has been useful in other sciences as well.

We have seen that the particle theory of light was supplanted by the wave theory. In the next few pages we will see that the wave theory itself has been outmoded. Some theories are replaced by better ones, just as an acorn is "replaced" by a mighty oak.

But some theories stand the test of time and remain intact despite all the new observations and experiments that are made. The tests verify the theory, rather than contradict it. Evolution is one of those powerful theories. Einstein's concept of relativity, as we shall soon see, is another.

11

The Ultimate Barrier

DESPITE NEWTON'S reputation, by the middle of the nineteenth century the wave theory of light reigned supreme. Wave theory described phenomena such as diffraction and refraction much better than the particle theory could.

Yet there was one aspect of the wave theory that was troublesome. If light is a wave, how does it get through the vacuum of space? Waves need a medium to propagate upon: Water waves propagate in water; sound waves in air or some other material.

In space, as the Hollywood ads say, no one can hear you scream. Well then, how can they see you in space if there is no medium for light waves to propagate in?

Hooke skirted this problem by saying that outer space is not a vacuum after all but is filled with an ether that serves as the medium for the propagation of light waves.

Curious stuff, this ether. It could not be seen. It could not be detected at all. It did not move; it merely filled the universe through which the Earth and all the other astronomical bodies moved. It was called the luminiferous ether (meaning "light-bearing element"). Its only function, apparently, was to allow light waves to move through space.

You might suspect that the ether was merely a figment of scientists' imaginations, invented to get around the inconvenient fact that light travels through a vacuum, which was supposed to be impossible. You would be right.

The luminiferous ether did not exist. It was shown to be nonexistent by two scientists who were intent on measuring the speed of light. Their work, known today as the Michelson-Morley experiment, was one of those crucial tests that destroyed a well-entrenched idea. It was a watershed that started physicists thinking in a new direction—including a sixteen-year-old named Albert Einstein.

A. A. Michelson did not start out to overthrow the concept of the luminiferous ether. All he wanted to do was to measure the speed of light more precisely than anyone had ever done before. "All." He was following in the illustrious footsteps of Roemer and a dozen other brilliant men, convinced that he could make a better measurement than they.

In 1881, with the financial help of Alexander Graham Bell, Michelson developed an *interferometer,* a device that split a light beam in two, sent the parts along different paths, and then brought them back together. If the two beams traveled different distances—or if they traveled at different speeds—when they were brought back together they would show bands of light and dark, an *interference pattern* such as Young had first studied nearly 80 years earlier.

Think of the two beams of light as two columns of marching soldiers. They start out in perfect step. But if one column marches a longer distance than the other, or happens to go a little faster or slower than the other, when they return to base they will no longer be in exact step.

The interferometer was a delightfully sensitive instrument. If the two beams of light fell out of step with one another so that their waves did not arrive on the final target at *precisely* the same instant, then the interference bands would show up. Excrutiatingly small differences could thus be measured quite precisely.

Together with Edward Williams Morley, Michelson determined to use the interferometer to detect the luminiferous ether. Their reasoning was basically simple: If the Earth is moving through the ether, then it should exert some sort of detectable effect. There ought to be an "ether wind" blowing across the surface of the Earth. It may be too tenuous for ordinary detectors to measure, but light waves should be sensitive to the ether wind.

One beam of light would be sent out in the direction of the Earth's orbital motion through space. The other would be sent in a direction perpendicular to the first one. If the ether existed, it should put up more resistance to the beam shining across the ether wind than to the beam shining in the same direction as the ether wind. The slightest resistance would show up as fringes of an interference pattern in the interferometer's detector.

No interference pattern was seen. No matter which way the interferometer was turned, the two beams of light traveled at precisely the same speed. There was no ether wind. The luminiferous ether did not exist.

It was the most spectacular negative result in the history of physics. Negative results are exceedingly valuable. It is crucial to know where the blind alleys are. The luminiferous ether was a blind alley.

More than that, the Michelson-Morley experiment, and Michelson's earlier interferometer work on measuring the speed of light, showed quite clearly that light traveled at the same speed no matter what other speed was added to it!

This was totally unexpected and the beginning of a true revolution in physics.

Look at it this way: If you are riding in a train and you get up from your seat and walk to the front of the coach, the speed of your motion—relative to the ground outside the train—is the sum of the train's forward velocity plus your own walking pace. For simplicity's sake we will say that the train is moving at 60 miles per hour and you walk at one mph; therefore your combined speed is 61 mph. But if you walk to the rear of the coach, contrary to the forward motion of the train itself, an observer clocking you from outside the train would see you moving at 59 mph—the train's forward motion minus your foot speed.

The Michelson-Morley experiments (and there were *many* measurements made over several years) showed that light does not behave this way. Shine a beam of light in the same direction as the Earth is moving and its speed (in vacuum) is 186,000 mps. Shine a beam of light in the opposite direction, so that the velocity of the Earth should work against the beam of light, and light speed is still the same!

This goes against common sense. But it turns out that this is the way the universe behaves. Light moves at the same speed no matter what.

But we are getting slightly ahead of the historical story.

Between 1864 and 1873, the great Scottish mathematical physicist James Clerk Maxwell produced one of those rare and precious bodies of thought that tied together all the known observations about electricity and magnetism into a single unified theory (there's that word again) of *electromagnetism*. Maxwell's theory showed that magnetism and electricity are one and the same force—the first time in the history of physics that two forces thought to be separate entities were proved to be different aspects of one and the same force.

(Almost exactly a century later, physicists would show that electromagnetism and the forces that govern the behavior of atomic nuclei are actually different manifestations of the same force. This unification follows and enlarges Maxwell's original unification of electricity and magnetism.)

It may seem strange that the force that allows a magnet to pick up a paper clip is part of the same entity as a lightning stroke or the current that lights the house lamps. It may even seem that we are wandering far afield from our interest in light.

But electricity and magnetism, Maxwell found, are intimately bound together. And light itself is a manifestation of this electromagnetic force.

Maxwell must be ranked among the handful of superstar physicists such as Newton and Einstein. His genius spanned a wide range of studies, including electromagnetism, light, color vision, astronomy, the behavior of gases, and the study of heat known as thermodynamics.

Maxwell was a mathematical physicist, a theoretician. His tools were pencil and paper. In his work on electromagnetism he found that electricity and magnetism are always found together. You cannot have an electrical current, for example, without a magnetic field being produced by it. And magnetic fields, conversely, are always associated with electrical charges.

Electromagnetic force travels as a wave, and electromagnetic waves have two components: an electric field and a magnetic field that vibrate at right angles to one another.

When Maxwell calculated the speed at which an electromagnetic wave travels in vacuum, the number came out to be roughly the same speed as light. Light's speed was not then known to the

James Clerk Maxwell, probably the greatest scientist of the nineteenth century, the man who first realized that light is a form of electromagnetic radiation.

precision it is today. Michelson did not even begin his measurements until about 20 years later. But the two numbers were very close to each other.

Maxwell jumped to a conclusion. He concluded that light is one form of electromagnetic wave. He had no real evidence for this, but the coincidence of that tremendous speed was too much to really be a coincidence.

On December 10, 1861, Maxwell wrote to a friend:

> I made out the equations [of electromagnetic force] in the country before I had any suspicion of the nearness between the two values of the velocity of propagation of magnetic effects and that of light, so that I think that the magnetic and luminiferous media are identical. . . .

Maxwell believed the luminiferous ether to be real, as did most scientists until the Michelson-Morley experiment proved otherwise. But his work in electromagnetism showed that electromagnetic waves could *propagate in a vacuum.* The logical conclusion was that light could propagate in a vacuum as well, and the luminiferous ether was not needed to explain how light waves could travel through empty space.

There were troubling aspects to this, especially the idea that some kinds of waves did not need a medium for propagation. What was happening, without anyone really knowing it, was that the serene and certain edifice of physics was being torn apart by new ideas, new observations, new experiments. For a while, in those calm Victorian years, most of the leaders of the field did not notice any threatening changes. The most eminent physicist of his time, Lord Kelvin, sadly told his students that the study of physics was just about finished. Everything that could be known had been discovered, he thought.

Like the report of Mark Twain's demise, Kelvin's melancholy was premature. The old edifice of knowledge that began with Newton was being demolished and a new structure was already starting to arise.

If Maxwell was correct, and light was a type of electromagnetic wave, then it should be possible to find other electromagnetic wavelengths that are not visible to our eyes. After all, we cannot see electric currents or magnetic force. In fact, *unless* such other wavelengths were discovered, Maxwell's hunch would be unproved and probably wrong.

A young German physicist, Heinrich Hertz, set up a simple experiment. His equipment produced an electrical spark between two metal balls, a sort of miniature stroke of lightning. Maxwell's equations predicted that such a spark should emit invisible electromagnetic waves. Across his laboratory, Hertz set a simple loop of wire, with a gap in it the same size as the gap of the spark source.

When Hertz allowed electric current to flow into the metal balls and cause a spark to flash, a similar spark immediately crackled in the wire loop across the laboratory room. Hertz had shown that invisible electromagnetic waves can travel over a distance from a transmitter of electromagnetic energy to a receiver, without wires or any other physical connection. The waves traveled at the speed of light; they were a form of electromagnetic energy of a wavelength much longer than visible light.

These invisible waves caused a sensation. They were quickly dubbed *Hertzian waves* and stirred enormous enthusiasm among physicists and engineers. Hertz had proved that Maxwell's surmise was correct. More than that, he had laid the foundation for what would eventually become the telecommunications industry. Hertzian waves today are called *radio waves*.

Hertz's experiment was performed in 1888. By 1895 Guglielmo Marconi was sending radio transmissions over the distance of a mile, and by 1901 Marconi succeeded in transmitting a radio signal across the Atlantic Ocean.

Seven years after Hertz's experiment, Wilhelm Roentgen stumbled onto the discovery of another form of electromagnetic energy. These waves could penetrate most forms of solid matter as easily as visible light goes through glass. Puzzled over their nature, Roentgen dubbed them X-rays, and that is what they are still called today. X-rays have, of course, been enormously useful in medicine, where they allow physicians to see inside the living body without cutting it open. They have also been used as a form of radiation therapy against cancer and skin diseases.

Today we know that the full electromagnetic spectrum ranges from intense gamma rays, which are emitted in the radioactive

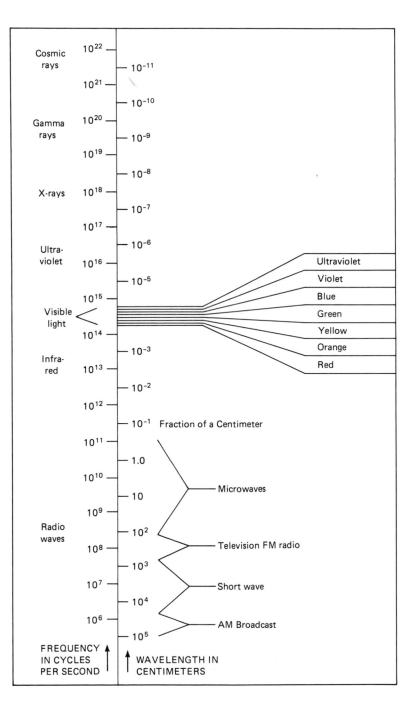

The full electromagnetic spectrum ranges from very long radio waves to extremely short gamma and cosmic rays. Visible light is only a small slice of the total spectrum of electromagnetic energy.

decay of atomic nuclei, to the kind of radio waves used in broadcasting. Gamma rays are of the order of 10^{-4} nanometers in wavelength, some 10 million times smaller than the wavelengths of visible light. AM radio waves range to some 10 million times larger than visible light's wavelengths; radio wavelengths can be miles long.

Hertz died at the tragically early age of 37 in 1894. But before he did, in 1889, he made a second discovery of great significance, although he never realized its true importance.

He found that when he placed a sheet of glass between his spark transmitter and the receiver, the spark in the receiver's gap became much weaker. Electromagnetic waves were still leaping across the lab, but somehow the pane of glass weakened them.

Hertz eventually concluded that the transmitter's spark emitted not only Hertzian waves but ultraviolet waves as well. When both kinds of waves traveled freely to the receiver, its spark was strong. When the glass pane blocked the UV, only the longer Hertzian waves got through and the spark was weaker.

His explanation was correct, but Hertz never realized that he had discovered an important phenomenon. It is *photoionization*, also known as the *photoelectric effect*. Hertz died before he realized its importance. Albert Einstein was awarded the Nobel Prize for his 1905 explanation of the photoelectric effect. Today, whenever you pass through an electric-eye door, you are seeing the photoelectric effect in action.

Simply put, some forms of electromagnetic energy are strong enough to knock electrons out of their orbits around atomic nuclei. Ordinary atoms have a certain number of electrons circling their nuclei. Electromagnetic energy of the proper wavelength can excite the electrons until one or more of them leave the atom altogether. The deprived atom is called an *ion;* where it was originally electrically neutral, it now has a net positive charge, since one or more of its negatively charged electrons have gone off and the positive electrical charges in the nucleus are no longer balanced.

The electrons thus freed can form an electric current, and that current can be put to use: For example, the current can operate an electrical motor that opens a door.

Voilà! Photoionization, or the photoelectric effect.

But how can waves of electromagnetic energy tear electrons loose from their atoms?

Right at the turn of the twentieth century, the German physicist Max Planck came to the conclusion that energy exists only in distinct packages, which he called *quanta*. Energy is not a continuous flow, like a stream of water. Energy comes in tiny lumps, in packets. A single packet is a *quantum*, and Planck's ideas were soon called the quantum theory.

Heresy! Planck seemed to be trying to revive the old particle theory of light, which everyone knew was wrong. Planck had been working in the field of thermodynamics, the study of heat and energy. But if his ideas were correct and all forms of energy are quantized, then electromagnetic energy—including light—must come in packets of quanta too.

But physics had progressed to a new sophistication. Planck's quanta were not like Newton's old microscopic bullets. Quanta can behave like particles, true enough. They can also behave like waves.

It was difficult for the established generation of physicists to accept the idea that light can be *both* particles and waves—and that the difference depended essentially on how you look at it. The quantum theory ran into powerful opposition. Yet it worked. It allowed physicists to understand things that could not be explained otherwise. It opened the door to new discoveries, new inventions. Sixty years after Planck's first enunciation of the quantum theory, the first laser was built.

Newton would not have been surprised by NASA's sending men to the Moon. He could have done the job himself if he had possessed the manufacturing capability to build large rockets. But the laser would have shocked Newton. It is based on principles that he never knew.

The work of Planck and the earlier discoveries of radioactivity and the photoelectric effect were the bombshells of a true revolution in physics. Young students found themselves in the midst of an intellectual ferment such as had not been known since Galileo was hauled before the Inquisition.

One of those students was a young German who had renounced his citizenship and gone to Switzerland to live and study.

"What would it be like," mused young Albert Einstein, "to ride on a beam of light?"

Many students have daydreams or pose questions that have no answers. Einstein found the answer to his question, and his answer has shaped the world we live in today.

In his book and television series *The Ascent of Man*, Jacob Bronowski pictured Einstein pondering this question as he rode the trolley car to work at the Swiss Patent Office in Bern during the first years of the twentieth century.

In 1905 his musings bore fruit. Einstein brought together Hertz's experimental work on the photoelectric effect and Planck's quantum theory in a short paper that eventually won him the No-

Albert Einstein, whose musings about the nature of light led to relativity, the quantum theory (which he abhorred), lasers, and Hiroshima.

bel Prize. It was one of three papers Einstein published that year. The other two dealt with Brownian motion—and the special theory of relativity.

Einstein's paper on the photoelectric effect explained how electromagnetic energy can strip electrons from atoms. The key to the phenomenon is not the wattage of the energy but its wavelength.

He showed that the wavelength of light (or any other form of electromagnetic energy) is associated with a certain amount of energy. Quanta of short-wavelength ultraviolet light, for example, carry more energy per packet than the longer-wavelength quanta of blue light. Blue light packs more energy per quantum than red, and so on. The more energetic the light, the more energetic the electrons that were stripped from the atoms.

In other words, Einstein found that the electrons absorbed the energy of the incoming light and carried it away with them. It's as if you were stuck at home because you did not have enough money to go out to the movies. Then someone handed you enough money to go off and see a flick. You zoom out of the house with the money in your pocket. Electrons stay in orbit until they receive the right amount of energy to liberate them, then they fly away from their parent atoms.

The ironic part of this story is that, although Einstein's work on the photoelectric effect won him the Nobel Prize and showed conclusively that Planck's idea of quantized bits of energy was correct, Einstein never accepted the quantum theory. He was opposed to it until the day he died.

The wave theory could not explain the photoelectric effect, nor many other phenomena that were then (pardon the expression) coming to light. Physicists began to call the basic quantum of electromagnetic energy the *photon*. Some aspects of the photon's behavior were best explained by describing it as a wave. Other aspects were best described as the actions of a particle. Half jokingly, physicists began calling the photon a *wavicle*.

The answer to the question of whether light is a stream of particles or a flow of waves is: Yes!

Or, as the poet Robert Frost put it:

We dance round in a ring and suppose,
But the Secret sits in the middle and knows.

Whether it is described as a wave or a particle, each photon has a certain definite amount of energy, no more and no less. This is called its *quantum energy*. The shorter the wavelength of the photon, the more quantum energy it contains.

In 1913 the Danish physicist Niels Bohr produced the basic explanation for the way that light interacts with atoms. His work showed how atoms produce photons and why the quantum theory correctly explains the nature of light.

Electrons can orbit the atomic nucleus at many different levels, like satellites orbiting the Earth at different altitudes. The electrons can change their orbits, going higher or lower, depending on how much energy they contain. An electron can absorb the energy of an incoming photon and jump to a higher orbit. Such an electron is said to be excited. Not just any amount of energy will excite an electron, however. Each electron orbit is susceptible to only a very narrow range of incoming energies. The photon must have the right wavelength or the electron will not absorb it and will not become excited.

Even when an electron does get excited, the excitement does not last long. In most cases the electron quickly jumps back to its original orbit—and gives off a photon *of precisely the same wavelength as it originally absorbed.*

Bohr showed the atom to be a dynamic thing, far more complex than the old Newtonian idea of a miniature solar system in which the electrons obediently circled the nucleus. The quantum-mechanical model of the atom pictures electrons hopping back and forth from one orbital level to another, absorbing incoming photons and emitting photons constantly.

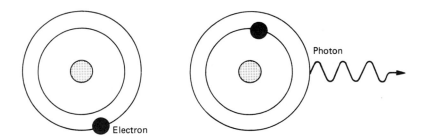

Electrons can orbit the atomic nucleus at different levels of energy. When a photon (a quantum of electromagnetic energy) is absorbed by the atom, the electron jumps to a higher orbit. When the electron gives off a photon, it falls back to a lower orbit.

Einstein's special theory of relativity showed that photons are the speed champions of the universe. No particle in nature travels faster than light. Einstein showed that no physical object—from subatomic particles to starships—*can* travel faster than light. More than that, Einstein's theory claimed that matter and energy are interchangeable.

The most famous equation in all of science is Einstein's $E = mc^2$. *E* is energy, *m* is the mass of the matter under consideration, and *c* is the symbol for the speed of light. The amount of energy inside any piece of matter is equal to the mass of that matter multiplied by the square of the speed of light, a huge number.

The atoms are storehouses of incredibly powerful amounts of energy, and if that energy could be liberated—Hiroshima was the result.

It is especially bitter that the gentle, pacifistic Einstein should be the father of the most awful weapon ever produced. Even as a teenager, Einstein hated the militaristic bombast of the kaiser's Germany so much that he became a Swiss citizen as soon as he could.

If Newton was adored by his colleagues and followers, Einstein was revered. He was often described as saintly. And, far more than any scientist of any generation, Einstein was made into a public figure. His special theory of relativity, which rocked physics, made him one of the most important scientists in the world. But the public did not know him until 1919, when a total eclipse of the Sun allowed astronomers to check a prediction he had made on the positions of certain stars. The prediction was based on his *general* theory of relativity, which he had published in 1915, ten years after the special theory.

The general theory stated that a very strong gravitational field, such as the gravity of the Sun, could bend beams of light. It had always been assumed that light traveled in straight lines; the idea of gravity bending light was new and unproved. The effect was very slight as well; it could not be tested easily. But the total eclipse of 1919 allowed astronomers (who had to travel to the jungle of northern Brazil) to observe the positions of stars that appeared so close to the rim of the Sun's disk that they could not be seen except when an eclipse blacked out the Sun's enormous glare.

The stars were where Einstein said they would be, and the man became the first scientist to be a media hero. "Newton's ideas overthrown," proclaimed the London *Times*. The general theory dealt with a new concept of gravity, an expansion of Newton's earlier work. But to the media, Newton was out and Einstein was in. In later years, as his genius and kindness became legend, he was revered by press, public, and even his fellow scientists. When Einstein left Europe to become the first professor at the newly established Institute for Advanced Study at Princeton, a colleague remarked that "the pope of physics" had moved to the United States.

If Einstein was canonized, two other German physicists were about to become notorious.

An awesome confirmation of Einstein's famous $E = mc^2$ *is the development of nuclear weapons. Einstein was so pacifistic that he renounced his native German citizenship before World War I to become a Swiss citizen.*

In the 1920s Werner Heisenberg and Erwin Schroedinger introduced what has come to be called the *uncertainty principle,* a logical consequence of quantum mechanics that has become badly misunderstood not merely by the public at large but also by science journalists and self-professed pundits of science.

Simply put, the uncertainty principle says that you cannot measure the position of a subatomic particle and its momentum (motion) at the same time. You can get a good fix on a particle's position, but then you won't be able to tell where it's heading. Or you can determine its momentum, but you won't know where it is.

This has been misinterpreted to mean that there is nothing certain about *anything.* People who should know better have wailed that the uncertainty principle proves the futility of life: You just can't know anything with any certainty at all.

Not so. The uncertainty principle deals with photons and subatomic particles, the kinds of particles that compose atoms. The uncertainty stems from the following.

To measure the position or momentum of anything, you must be able to see it or detect it in some manner. Whales measure the position of a school of fish by bouncing sound waves off them, like sonar. Dogs find buried bones by detecting odor particles seeping through the ground. Some species of shark detect their prey by sensing the electric field the prey emits.

When you or I measure something—say, the distance to a chair across the room—we do it by sight. Most of our measuring tools are vision oriented, from tape measures to microscopic calibrators. They involve bouncing photons off something and receiving those photons in our eyes. A common adage that physicists tell one another is, "All of physics comes down to reading a number on a gauge."

When we get down to the sizes of subatomic particles, they are so tiny that the photons push them around. Send in a photon to measure the position of a subatomic particle and it actually pushes the little fellow as it bounces off it. When the photon gets back to your receiver it can tell you where the particle *was* when it struck the little guy; it cannot tell you where it is now.

That is the essence of the uncertainty principle. It has nothing to do (except in the broadest philosophical sense) with everyday affairs. You can still measure the height of the Eiffel Tower down to teeny fractions of a millimeter. You can still find the distance to the North Star. You can still believe in love and faith. The world is not going to dissolve into a puddle of Heisenbergian uncertainty all around you.

By the middle of the twentieth century, physicists had answered the question, What is light?

Light is a form of electromagnetic energy. Light is a narrow slice of wavelengths in the very wide spectrum of electromagnetic energy. Other wavelengths have given us radio, television, and X-rays for medical diagnosis and therapy. The quest to learn light's true nature led Einstein to the special theory of relativity, which eventually gave us nuclear power, and in 1915 to the general theory of relativity, which supplanted Newton's concept of gravity.

The English theoretician P. A. M. Dirac boldly proclaimed that the new quantum physics "explains all of chemistry and most of physics." It brought us the modern telecommunications industry, CAT scanners, radio telescopes, nuclear power plants, transistors, lasers, electronic computers, and a host of other new capabilities.

It also brought us the hydrogen bomb and missiles that can strike across intercontinental distances in half an hour, transistorized "boom boxes" carried by youths that shatter the environment around them with deafening noise, puerile sit-coms on TV, and a national attention span geared to 30-second commercial messages.

All in all, though, the new knowledge gained by twentieth-century physics has helped to make our lives longer, healthier, and richer. This new knowledge has also shown us that there are limits to our abilities, ultimate barriers that apparently cannot be crossed.

The uncertainty principle has shown that there is a fundamental limit to our ability to measure the physical world. The special theory of relativity has shown that there is a fundamental speed limit in the universe: the speed of light. Nothing can travel faster than light.

Limits, yes. But the new capabilities that the physicists have opened for us are enough to allow us to expand our domain down to the basic bits of matter that hold the atoms together and out to the farthest galaxies and quasars. The quest for understanding the true nature of light has allowed us to comprehend the universe, from its smallest constituents to its widest grandeur.

III.
TO
USE

[Art is] anything done so superlatively well that we all but forget what the work is supposed to be, for sheer admiration of the way it is done.

—*E. H. Gombrich*

. . . it is more important to have beauty in one's equations than to have them fit experiment.

—*P. A. M. Dirac*

12

"I'll Do Your Looking for You"

IN THE CLIMACTIC scene of Ray Bradbury's motion picture script for Herman Melville's classic novel *Moby Dick,* Captain Ahab is at the bow of a longboat, chasing the white whale. Cunning and dangerous, Moby Dick dives deep. The rowers peer anxiously about them, looking for signs of the dreaded killer. Ahab commands his men to concentrate on their oars and not to look up.

"I'll do your looking for you," says Ahab.

In a sense, that is what great artists do. They do our looking for us. They perceive unities and beauties that we miss. Their talented eyes and minds see the world in better detail than most of us do, and their trained hands reproduce a kind of beauty that we would never have noticed.

Great visual art, be it painting or sculpture or photography, begins with the eyes. All the great artists knew this.

This chapter deals with the way artists have used light in their works. I will concentrate mainly on painting because it is the art form that delights me most and because the painters faced not only the problems of form and color but the unique problem of trying to present a convincing picture of the three-dimensional world on a flat, two-dimensional surface. Over the centuries the painters succeeded so well in solving these problems that nowadays they

153

have gone beyond pictorial representation of the world we can see and into abstractions that have little connection with the physical world around us.

The visual arts began in the snows and cold of the Ice Age, when early humans first started to decorate their world. (See Plate 10 following page 174.) They shaped stones into rough figures of animals and pregnant women. They scratched primitive drawings onto the animal bones they used as tools. Fossilized bones and sticks have been found to bear markings that were not pictures but marks that probably represent the first attempts at counting and arithmetic. Art and science were never far from one another in those early days. As we will see, even as late as the Renaissance, art and science progressed hand in hand. It took our modern pseudosophistication to separate the two.

Painting has been an important art form for at least 30,000 years. The earliest evidence comes from cave dwellers who drew the outlines of their hands on rocks and colored them in. They drew meandering lines of charcoal inside their caves. Decoration? The innate human desire to "make one's mark"? The earliest graffiti?

Cave-dwelling Cro-Magnon hunters painted hauntingly graceful pictures of bison, horses, lions, gazelles, and even animals that are now extinct, such as the aurochs and woolly mammoth, on the walls of their caverns. These lovely and powerful paintings, which have been dated at some 20,000 to 30,000 years old, have been found chiefly in the southwestern portion of France and northeastern Spain, where the climate has allowed them to survive. But natural disasters are not the only threat to such treasures. Caves such as Lascaux and Altamira have been closed to the public to protect the paintings against the ravages of modern tourism.

The colors these early artists used came from the materials around them. Soil rich in iron oxides provided reds and browns. Manganese oxide produced blue, iron carbonate yielded yellow. Charred wood from their fires gave them black. Chalk or ground seashells made white. Later on, as the ice retreated and vegetation became more abundant, crushed berries, roots, and leaves yielded a broader palette of colors.

The cave dwellers went to considerable effort to produce their art. The pictures they drew were not spur-of-the-moment doodles. They developed tools and paints purposefully and must have devoted much time and trouble to make their paintings.

Cro-Magnon artists ground the natural pigments they found into powders and mixed them with water. In some locations they heated the pigments, which changed the natural colors into new tones. Naturally occurring magnetite, for example, is black. But when heated it combines with oxygen from the air and turns into red hematite.

Archaeologists have found painting utensils at the cave sites: hollow bone receptacles (such as skulls) for holding the paints, bone mortars and pestles for grinding the pigments, even stone lamps for lighting the dark recesses of the caves.

By the time of the Middle Ages, the artist's palette had gained a much richer variety of colors, although their sources were still much the same as before: natural pigments from colored earths and vegetable squeezings.

In medieval Europe a different art form arose, one in which the artists literally painted with light. I refer to the incredibly beautiful stained-glass windows of the great Gothic cathedrals. (See Plate 11 following page 174.)

Cathedrals such as those in Chartres, Cologne, and Amiens, and Notre Dame in Paris are among the finest architectural feats of the human mind. The techniques of Gothic architecture allowed stone structures to be built as delicately as lacework and to soar

Our cave-dwelling ancestors created beautiful works of art with primitive tools and pigments.

heavenward as no buildings could before. The weight of these magnificent edifices rested on stone pillars and buttresses; walls were not needed to hold the structure together. Solid stone could be replaced by windows to bring light into the church.

And what windows! In the twelfth and early thirteenth centuries unknown European glassmakers and artists produced stained-glass windows of breathtaking beauty. Nothing like them has been produced since, although stained glass has enjoyed several revivals in popularity up to our own age.

The glass made in those medieval decades was crude by comparison to modern glassmaking, yet its very crudeness added to the artistic effect of the finished windows. The glassmakers could not produce subtle colors. They added metallic oxides to their glass while it was in the molten state: copper for ruby red, cobalt for blue, manganese for purple, antimony for yellow, iron for green. The result was pure, rich, deep, bold colors that still strike the eye today with power and glory. The glass itself was uneven in thickness, rough by today's refined standards. That gave it a texture and depth of color; it was what artists call a happy accident.

What sets stained glass aside from other forms of visual art is that light itself orchestrates the finished work. A stained-glass window is not a static picture. As the day progresses, the different angles and intensities of sunlight illuminate and alter the window, charging the vibrant colors with energy, toning them down as the hours pass. To see these masterworks in all their splendor, you should spend an entire day in one of the magnificent cathedrals and watch how the shifting sunlight interacts with the colors and forms of the glass.

In a sense, the art of stained glass suffered from growing refinements. The artists learned how to make smoother glass and subtler colors. The power and glory of the Age of Faith gave way to new interests, new conflicts. Over the centuries artists have often returned to the medium of stained glass but never with the simplicity and sheer visual impact of the earlier works. About a century ago Louis Comfort Tiffany produced stained-glass lampshades and other household objects that are much prized by antique hunters. (See Plate 12 following page 174.)

Yet the stained-glass windows of the great Gothic cathedrals stand alone in their power and beauty to this day.

Over the centuries, artists and artisans have developed rich palettes of colors for the purposes of painting, printing, dying, and otherwise beautifying the objects that we use every day of our lives.

Generally speaking, there are two types of coloring agents: dyes and paint pigments. Paint pigments are usually powders that are suspended in a medium such as oil. When paint is applied to a surface, the oil dries and hardens, leaving the microscopic particles of pigment scattered throughout it. The colors that we see come from light that strikes the paint and is scattered back toward our eyes by the pigment particles.

When I was in kindergarten at the Abraham S. Jenks elementary school in south Philadelphia, we were given watercolors to paint with. I watched the teacher mix red and blue to produce purple. I watched her mix red and yellow to produce orange. Yet when I mixed the colors from my little box of paints I usually got nothing but a drab gray. Once in a while I could achieve a fairly ugly shade of brown. Watercolors defeated me.

Only much later in life did I learn that the main reason we were given watercolors was that they were cheap. Ours was not a well-to-do school district. No telling how many potential Rembrandts were turned off painting in kindergarten by the watercolor experience. To this day, although I can sketch cartoons as well as most doodlers, the thought of *painting* a picture gives me the shudders.

Watercolors consist of a pigment mixed with an adhesive substance. When you add water, it forms a colored liquid that can be spread on a surface. The surface should be high-grade paper or plaster—neither of which was provided at dear old A. S. Jenks. When you apply watercolor to paper the water quickly evaporates, leaving the pigment attached nicely to the paper. But there is no protection for the pigment. If you get it wet again it will flow again. That was my big problem. My blue sky kept bleeding into my green meadow. The result: gray on gray.

About 2000 B.C. the Egyptians figured out a better way. Tempera paint added the yolk of an egg and a few other ingredients to the pigment and produced a hardy product that was easy to apply and dried quickly to a tough finish. The egg yolk, a mixture of a fatty oil, sticky albumen, and water, served to dissolve the pigment into very fine particles. Its faint yellowish color soon evaporated, leaving the pigment nicely embedded in the dried albumen. And it stuck to the surface being painted like dried egg yolk sticks to a dish.

By the fifteenth century oil paint became the dominant medium for artists. The pigment is mixed with (usually) linseed oil. While the Flemish brothers Van Eyck are often credited with inventing oil painting, the true origin of oil paints is undoubtedly centuries older. Oil paints are much easier to use than tempera (I will remain discreetly silent about watercolors), although the hardened oil does tend to darken over years of time. Many of Rembrandt's paintings, for example, turned out to be much brighter and richer than thought once they had been cleaned of the oil film that had darkened over them.

Modern acrylic and latex paints use a plastic binder in place of oil or egg yolk. Plastic binders are easy to use and quite durable. However, some artists believe they lack the subtleness of color that is possible with oils.

While paint pigments are suspended in a binder such as oil or plastic, *dyes* are color materials that are usually dissolved in a solvent. Dyes can be dissolved in gelatin or clear plastic to make the colored filters used in stage lighting and photography. Color film is based on such dyes. Dyes that are soluble in water or other solvents are used for coloring fabrics and paper. (See Plate 13 following page 174.)

There is actually a third coloring agent called *lakes,* which are pigments that consist of particles that have been dyed.

Dying has been an industry since before the beginning of written history. There was money to be made from producing colored cloth. The Chinese were wearing dyed clothing as early as 3000 B.C. By 2500 B.C. India was producing red dye from the root of the madder plant and blue from indigo plants. The Egyptians were also dying their linen clothing in saffron yellow from dried crocuses, as well as greens and reds.

By the time of the Roman Empire, the eastern Mediterranean city of Tyre had a monopoly on a dye that came to be known as royal purple. Obtained from a certain species of sea snail and other shellfish, Tyrian purple was so expensive that only the very wealthiest and most powerful people could afford it. Thus it came to be associated with royalty. Only the ruling Roman families could wear togas edged with Tyrian purple. To this day, we associate the color with royalty.

Dyes were produced from natural sources down through all the centuries until 1856, when a 17-year-old English chemistry student, William Perkin, discovered that a substance obtained from the black sludge of coal tar could, when mixed with alcohol, dye silk purple. This first synthetic dye was colorfast, too, meaning that it did not wash out easily.

Perkin did what any modern student would do. He dropped out of school, borrowed money from his father, and developed the first artificial dye process. He became very rich, was knighted, and before he died (at the age of 69) he developed the first artificial perfume, also from coal tar.

Perkin's dye was the first of the so-called aniline dyes, aniline being the stuff from the coal tar from which the dye was derived. He called his dye aniline purple, but the more romantic French called it *mauve,* after their word for the purplish mallow flower. Mauve struck the haut monde world of Europe like a thunderbolt, and soon no fashionable lady would be seen without the color adorning her ensemble. The Mauve Decade symbolized the late nineteenth-century Belle Epoque world of Paris.

More than that, though, Perkin's work opened the door to the entire organic chemistry industry—which eventually led to plastics, pharmaceuticals, fertilizers, and more.

Today dyes are still made principally from coal tar products or from petrochemicals, which come from oil.

And good old indigo dye, one of the earliest known to humankind, is still very much with us. It is the dye that makes blue jeans blue. However, if the jeans were completely dyed with indigo they would look more like a pair of walking neon bright lights than a pair of trousers. They would be garishly, embarrassingly blue. So blue jeans are only half dyed. The vertical warp threads are dyed with indigo, but the horizontal weft threads are left white. The result: Blue jeans are popular around the world, especially when they've been washed or otherwise treated to tone down their original brightness even further.

But we have gotten ahead of ourselves. Back to painting.

Over the centuries, artists have acquired an empirical, trial-and-error knowledge of color, its uses, and how human beings perceive colors. The scientific study of color perception began only a scant few centuries ago. It is no accident of history that art and science flowered together during the Renaissance in Europe. As artists strove to understand the nature of color and how the human mind perceives colors, as they struggled to produce paintings on flat surfaces that gave the appearance of fully three-dimensional figures, they necessarily studied the physical nature of the eye and the brain and the physiology of vision. (See Plate 14 following page 174.)

Nor do the arts flourish separately. Painting, sculpture, and architecture seem to flower together at various periods of history.

Primitive men drew beautiful colored pictures on the walls of their caves. They also carved figures from bone and rock. And decorated their earthenware vases and pottery with colorful designs. They produced art, which art historian E. H. Gombrich defines as "anything done so superlatively well that we all but forget what the work is supposed to be, for sheer admiration of the way it is done."

Over the millennia that followed, sculpture seemed to advance beyond painting. Sculptors were able to produce works of enormous beauty, power, and realism. The monumental statuary of ancient Egypt still awes the newcomer with its serene strength and majesty. The lifelike statues of Phidias and the other, unknown sculptors of ancient Athens are still marvels of grace and beauty. (Incidentally, they were never intended to be seen as they are today. The Athenians painted their sculptures; the gods' beards were black, the goddesses' robes were gold, their eyes were painted in, and perhaps even their skins were flesh-toned.)

But the painters of pictures could not portray fully rounded three-dimensional figures on their flat boards and walls. Not even the best of the ancient painters could draw in perspective, as any grammar school student is taught to do today. No one knew how.

The laws of perspective were developed in the early Renaissance, when new interest in the heritage of knowledge left by the Greeks and Romans stirred European thinkers into new attitudes and discoveries.

Leonardo da Vinci is often credited with developing the laws of linear perspective, but he was not alone in this. Filippo

Brunelleschi, the great Florentine architect who died six years before Leonardo's birth, is the true father of linear perspective. In *The Story of Art,* Gombrich writes: "It was Brunelleschi who gave the artists the mathematical means of solving this problem [of perspective]; and the excitement which this caused among his painter-friends must have been immense."

While Euclid, around 300 B.C., understood that the size of an object as seen by the eye depends both on its actual size and its distance, it was not for another 16 centuries that Europeans developed the mathematics and did the physical experiments that solidified the laws of perspective. Artists developed an apparatus that has come to be called a *Leonardo box,* a device for physically showing how an object should be drawn from various points of view.

Renaissance artists developed the techniques of perspective using tools such as the "Leonardo box." While Leonardo da Vinci got the credit, the true father of linear perspective was Filippo Brunelleschi.

The Leonardo box allowed artists to learn how to draw an object that is foreshortened, that is, distorted because of the angle at which it is viewed. They would place the object—say, a lute—on a surface and then connect strings from various parts of it, through a frame that represented the picture's frame, to a focal point that represented where the eye of the beholder would be. Then they measured exactly where each string passed the plane of the frame; so many inches from the top, so many inches from the left side (or the bottom and the right side). A dot could be placed on a sheet of paper to represent where that string crossed the plane. With enough strings, and enough dots, the figure of the lute took shape on the paper as it would be seen by the eye viewing it from that angle.

Suddenly Florentines were gaping at pictures of such solidity and reality that they found it hard to believe they had been painted on a flat wall. Masaccio's *Holy Trinity,* in the church of Santa Maria Novella in Florence, seemed to them as if a hole had been broken through the wall of the church to reveal the Crucifixion scene. The figures were as solid and three-dimensional as a sculpture. (See Plate 15 following page 174.)

Perspective did indeed cause immense excitement among the painters. According to Gombrich, Paolo Ucello, a contemporary of Brunelleschi:

> . . . spent nights and days drawing objects in foreshortening, and setting himself ever new problems. . . . He was so engrossed in these studies that he would hardly look up when his wife called him to go to bed, and would exclaim, "What a sweet thing perspective is!"

Thus it was no accident of history that mathematics and the laws of linear perspective developed simultaneously in the Renaissance. The burgeoning of curiosity and invention brought new knowledge of the arts and the sciences simultaneously.

Leonardo, that troubled genius who was born 40 years before Columbus discovered the New World, is the epitome of the Renaissance artist, curious about all things, delving into human anatomy despite the prohibitions of the Church because *he had to know* how the body was built and how it worked in order to paint it properly. He had to know how human vision worked in order to create pictures that showed what he wanted his audiences to see.

To this day, his *Mona Lisa* is the best-known single painting on Earth. There is more than artistic genius behind Lisa Giaconda's smile. There is the early prescientific understanding of color con-

The culmination of Renaissance painting and the most famous single work of art on Earth: Leonardo's La Giaconda, *better known as the* Mona Lisa.

trasts, of perspective, of shading, and of subtle psychological tricks such as the fantasy landscape in the background of the picture.

Leonardo worked alone. His voluminous notes were written in code and backward, so that they could only be read in a mirror. He was not a scientist in the sense of Newton because he did not share his knowledge with the world. He may have been interested in learning about nature, but he kept his knowledge to himself—and showed the outside world only the results: paintings such as the magnificent *Last Supper,* and mechanical contrivances that have been lost over the course of the years. He was a solitary genius, and his secretive notebooks show that his imagination ranged far beyond the limits of his time.

But science is a collaborative endeavor. It is a cathedral that is constantly being enlarged and strengthened, stone by stone, by each generation of workers. Science cannot advance in secrecy. Leonardo might have been the most brilliant scientist of them all, had he shared his thoughts more fully. As it happened, the *organized* investigation of light and color had to wait more than a hundred years after his death.

When scientists such as Newton and, later, Maxwell probed the nature of light and the basis for color, they eventually saw the connections between the phenomena of the world around us and the workings of the visual system inside our heads.

We have seen that the three types of cones in the retina are sensitive to red, blue, and green light. Scientific experiments, beginning around the time of Isaac Newton, laid a firm foundation of understanding about color vision and color mixing.

Each color that we perceive consists of light of certain wavelengths. For example, blue light is concentrated in the wavelengths between 450 and 500 nanometers; green, about 525 to 575 nm; red, from roughly 650 nm out to beyond human perceptual range. It is possible to measure the actual power, in watts, of a beam of light by using a *spectroscope,* an instrument that can break up white light into its component colors, like a prism, and detect not only the wavelengths of incoming light but also their energy content.

The spectroscope allows scientists to produce *spectral power distribution* (SPD) curves, graphs that plot the energy content of various colors against their wavelengths. Thus the spectroscope

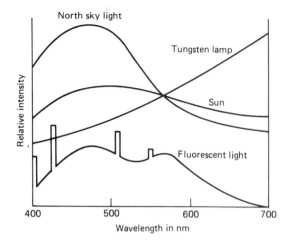

Spectral distribution curves plot the energy content of light (relative intensity) against the wavelengths (in nanometers). The sources of light plotted here include the northern sky, a common tungsten electric lamp, the Sun, and a fluorescent lamp.

allows us to measure the physical properties of colors. From this knowledge we can predict how these colors will be perceived in the brain and how they can be mixed to produce all the various hues of the rainbow. What painters did by hard-won trial and error, science has organized into a formula.

Though it is impossible to describe a color subjectively, colors can at least be defined objectively in terms of their wavelengths and SPD. As we saw earlier, you would have a hard time telling a friend over the telephone exactly what color your new coat is. But if you had access to a spectroscope, you could tell your friend the wavelengths of the coat's color and the energy content of the light reflected from it. That might help your friend visualize its color—if your friend were a physicist.

Our perception of colors depends not only on the wavelength and energy content of the light our eyes receive but on two other attributes as well. Psychological studies have shown that *saturation* and *brightness* have important effects on how the brain perceives the colors that the eye sees.

Colors come in various *hues,* or wavelengths. In addition to its hue, though, a color has the property of saturation. In this sense, saturation means the lack of "whiteness" in the color, how much the color differs from white or gray. Pink, for example, is unsaturated red. If you are mixing paints, to make pink you would blend red with white. Harvard's crimson is a highly saturated red; no need for blending white with it.

Brightness refers to the intensity of the color. Day-Glo orange, the color that school crossing guards adorn their uniforms with, can be seen by even the most myopic motorist from a good distance away. It is a very intense, very bright color. The noonday Sun is a very bright yellow, too bright to look at for more than a fleeting fraction of a second. When that same Sun has made its way to the western horizon, not only has its hue changed (due to atmospheric absorption) but its brightness has gone down.

Brightness has an effect on our perception of color. Increasing brightness makes reds appear more yellow and violets appear bluer. Colors at the red end of the spectrum, that is, with wavelengths longer than about 575 nanometers (green), appear to shift toward the blue as brightness increases; colors at the blue, short-wavelength end of the spectrum seem to shift toward the red. With decreasing illumination they shift the other way. In other words, as brightness increases colors seem to shift toward the middle range of the visible spectrum. As brightness decreases, reds become redder and blues bluer.

White light, however, remains pretty constant no matter the brightness of the illumination. As we shall shortly see, this *brightness constancy* forms the basis for one of the theories of color vision.

Colors produce emotional reactions in us. We tend to think of reddish colors as warm and of bluish colors as cool. Red, yellow, and orange are usually associated with excitement, passion. Blue and green are restful, calming hues. Purple is often identified with dignity or even sadness, while black, brown, and gray are the colors of melancholy and depression.

Our emotional responses to colors have given phrases to our everyday speech. We see red, feel blue, become green with envy, purple with rage, or sink into a brown study. A red light means Stop! Danger! A green light is the go-ahead signal. We associate black with death, white with purity, and bloody red with anger.

Physiological studies have shown that colors can influence blood pressure, heartbeat, breathing rate, perspiration, and even brain wave patterns. However, these physical responses are usually transitory. And repeated exposure to the same color does not produce a repeated physical response. Familiarity breeds not contempt but ennui.

Newton and those who followed him brought organized sense to the rules of thumb that artists had used for centuries. In particular, the scientists uncovered the physical basis for color mixing.

Remember the problems I had with watercolors? Newton was smart enough to work with light, not water paints. Using monochromatic beams of light, individual colors coming out of his prisms, he found that if he mixed green and red light he obtained yellow. Mixing blue and red produced magenta (purple).

Newton had hit upon the technique of *additive color mixing.* That is, one light is added to another. The two different spectral power distributions add together to form an SPD that is the sum of the two original SPDs. Not only does the color change but the brightness is increased as well. (See Plate 16 following page 174.)

Young showed that virtually all the colors we can perceive can be created by blends of blue, green, and red. Today we realize that these are the colors that the cones of our retinas are sensitive to. By mixing these *primary colors,* almost any other color can be produced. Mix all three primary colors together and what do you get? White.

If two colors produce white when they are added together, they are called *complementary.* The complement of a primary color is called a secondary color.

To illustrate, we saw a few paragraphs above that mixing blue and red yields magenta:

$$B + R \rightarrow M$$

Blue plus green produces turquoise, more properly called cyan:

$$B + G \rightarrow C$$

And mixing the three primaries—blue, green, and red—produces white:

$$B + G + R \rightarrow W$$

Now then, if the complement of a color is the color that will produce white when it is added to the original, then the complement of red is cyan, since cyan is composed of blue plus green, and red mixed with blue and green yields white:

$$R + C \rightarrow W$$

Cyan, then, is a secondary color and the complement of red.

To make sense out of additive color mixing, and simplify the rules, Newton devised the first *color wheel*, a chart that shows the primary colors arranged in a circle around "basic white," with the complementary colors placed diagonally opposite the primaries. (See Plate 17 following page 174.)

All well and good, as far as the colors of light are concerned. But what about my water paints?

What my kindergarten teacher did not tell me (possibly because she did not know) is that mixing paints is very different from mixing lights. When it comes to mixing paints we enter the realm of *subtractive color mixing*. (See Plate 18 following page 174.)

This is easier to understand if we begin with rays of light. Subtractive color mixing takes place when we pass a beam of light through a colored filter. The filter can be a piece of colored glass or plastic, even a container of liquid that is colored by a dye. The filter absorbs part of the light, makes it dimmer when it emerges on the other side.

It also changes the light's color. The filter absorbs some of the wavelengths of the incoming light. It alters the light's SPD. In a sense, the filter plays the same role as a university's entrance examinations. All sorts of students want to attend the university; they take the exam. Only those with the necessary knowledge get past the exam and actually enter the university. They are "on the right wavelength" to become university students. The others are not allowed into the school. Similarly, a colored filter stops (absorbs) all the light that is not "on the right wavelength" to pass through it.

Filters are usually called by the color of the light they allow through them. A red filter, for example, absorbs all the wavelengths of white light except those down at the long end of the visible spectrum. A blue filter stops everything but the shortest wavelengths.

Take a beam of white light and pass it through two filters, one red and one blue. What comes through the second filter? Nothing. Nothing visible, that is. All the visible wavelengths have been absorbed by the two filters.

Subtractive color mixing is trickier than additive because you can use more than one filter. You can start with a beam of white light, for instance, and pass it through a magenta filter and then a yellow one. The magenta filter allows only blue and red to pass through. The yellow filter, by itself, would pass green and red, but since the magenta filter has already screened out the green light, only red gets through the yellow filter.

For subtractive color mixing, the primary colors are magenta, yellow, and cyan. Combinations of two or all three of these colors can produce all the colors of the spectrum out of white light. Recognize that magenta is actually a combination of blue and red; yellow is a combination of green and red; and cyan is a combination of blue and green. Magenta, yellow, and cyan are each products of the three basic colors that our retinas are sensitive to.

Notice, too, that magenta, yellow, and cyan are the complements of the three additive primaries: blue, green, and red.

This is more than arcane trivia for physicists. The rules of subtractive color mixing govern any and all processes involving colored materials—such as paints and inks. When Leonardo painted the *Mona Lisa,* and when Hugh Heffner pored over the proofs of *Playboy's* latest centerfold, the colors they saw depended on subtractive color mixing.

The colors in paints and inks are caused by pigments suspended in the fluid. Many pigments exist in nature. As we have seen, our ability to see colors depends on pigments in the cones of our eyes. Our ability to see *at all* depends on the pigments in our retinal rods and cones.

In printing, the pigment of the ink attaches chemically to the paper while the fluid portion of the ink evaporates. Touch the paper too soon after printing and the ink smears; you must wait for it to dry for the fluid to evaporate and the pigment's chemical bonding to the paper to take hold.

In oil paints, the fluid is usually linseed oil, which dries on the canvas with the pigment particles and helps to provide a protective coat for them. The particles of pigment are suspended in the oil even after it dries; they do not lie flat on the canvas. Instead, they are "frozen" into the dried oil, suspended like raisins in a cake. When light strikes the painting, it is reflected off these suspended particles of pigment. This gives the feeling of texture to a painting, because the light striking the paint can be reflected into our eyes from various depths in the layers of suspended pigments. Great artists can make you almost *feel* the fabrics that the people are wearing, the glossiness of the glass the lady is holding, the softness of the little dog's fur.

The light reflected from the paint particles takes on the color of the pigment, following the laws of subtractive color mixing. Blue pigment, for example, absorbs all the incident white light except the blue wavelengths—those it reflects into our eyes.

Each pigment, therefore, absorbs some light and reflects some. The pigments act in much the same way as any filter, although now the light is bouncing off the pigment particles rather than passing through them. Every substance in the universe absorbs or reflects light; even the purest Baccarat crystal bounces a teeny portion of light off its surface while letting the rest of the light through. Nothing is totally transparent.

Scientists study the way light is reflected by objects and draw up *reflectance curves* that show which wavelengths, and what percentage of the incoming light, are reflected. Green plants absorb red and blue light pretty effectively and reflect light from the middle ranges of the spectrum; that is why they appear green.

Now we can see why fluorescent lights can make even the loveliest complexion look ghastly. The colors we perceive depend on the spectral power distribution of the light *and* the reflectance of the lady's skin. Even though the reflectance curve of the skin stays the same, if the SPD of the illuminating light changes, our perception of the lady's complexion will change too.

Modern four-color printing is based on the subtractive primaries of magenta, yellow, and cyan, plus black. These colors are sprayed onto the paper as microscopic dots that partially overlap. Where they overlap, subtractive color mixing determines the color seen by our eyes. Where they do not overlap, the colors are determined by a modified form of additive color mixing.

Additive and subtractive color mixing determine how we perceive colors. They are the basis of all painting and printing. In turn, color perception depends on the hue, saturation, and brightness of the colors we see. But there are other effects involved in color perception that are due to the "wiring" of the photoreceptors and nerve cells of the retina.

By the time the Renaissance was burgeoning in Europe, painters realized that they could play certain tricks on the eyes of their

beholders. For example, the colors we perceive depend to a considerable extent on the brightness and saturation of the colors surrounding the one we are looking at. Moreover, the eye can be fooled by optical illusions that make two objects of the same size appear to be different sizes.

Most of these tricks depend on the phenomenon of *contrast*. The brain tends to judge colors and sizes, at least in part, by comparing them to the other objects in view. Hollywood learned this in the early years of the motion picture industry. When you want the hero of your western to be a strapping six-footer but the

The eye sees, but the brain perceives. Specially designed settings can trick the brain into perceiving an optical illusion.

actor playing the role is only five-three, you don't stretch the actor, you lower the door frames. Western sets are built small so that the actors will look bigger. A standing joke around the studios is that John Wayne was only five-foot-six and rode a Shetland pony. (The Duke was actually well over six feet tall; no one ever had to downsize a set for him.)

You know from experience that your vision can adapt to low levels of light very quickly. Step into a movie theater that is so dark you can't see the seats at all, and in a few moments your eyes will adapt and you will be able to find an empty seat without trouble.

Your vision can adapt just as quickly to increases in brightness. From time to time I am interviewed by television news reporters. Almost invariably, when the camera crew first turns on their powerful lights my eyes squint and I feel something very close

Optical illusions exist in nature, also. Light reflected off a dry surface gives the illusion of water. Such mirages are common on highways, as well as in the desert.

to physical pain. Yet in a few moments my eyes adjust and I can conduct the interview quite easily.

This is called *lateral inhibition.* The eye adjusts to increases in illumination in two main ways. First, the iris closes down, reducing the amount of light allowed into the eye. This can cut down the incoming light by a factor of 16. Second, the retina's sensitivity to light is automatically regulated by the chemical reactions of the pigments in the rods and, in the case of bright light, particularly in the cones.

The sensitivity of one area of the retina is modified to some extent by the intensity of light falling on the retinal areas around it. This is called *lateral brightness adaptation.* For example, an area of gray will appear lighter when surrounded by white than it will when surrounded by black.

Such effects also apply to our perception of colors. Goethe's *Theory of Colors,* published in 1810, gained notoriety for its attack on Newton's work. Goethe cited many examples of color-contrast phenomena that he believed refuted Newton. They did not, but the examples did pose fascinating questions about how colors are perceived. One of Goethe's examples was to light a candle at twilight and place it on a white sheet of paper, then hold a pencil upright so that the candle cast its shadow upon the paper. The shadow, with the dying rays of the Sun falling on it, will appear a beautiful blue.

Explain that! Goethe challenged. Newton did not (he was long dead), but later scientists did.

It is a matter of lateral inhibition. The yellow candlelight stimulates the retina's red- and green-sensitive cones preferentially. The shadow of the pencil is illuminated by white sunlight. But the sensitivity of the red and green cones has been reduced by lateral inhibition; it's as if they are saying to the brain, "Hey, I'm already busy taking in this candlelight. I can't handle more work right now." What is left is the blue-sensitive cones; they take over the chore of sending information about the pencil's shadow to the visual cortex, but the information is colored blue, because that is the signal those particular cones send.

The basic theory of color vision was laid down by Young nearly two centuries ago and has been quantified by later scientists. Young's *trichromatic* theory, the concept that all the colors we

The brain can "fill in the blanks" to make a comprehensible picture out of a disjointed series of shapes, under certain conditions. Psychologists have used ink-blot pictures to probe the mental perceptions and attitudes of patients.

PLATE 10

In the Ice Age snow and cold of 30,000 years ago, Cro-Magnon artists used natural pigments to create breathtaking paintings, such as this "sorcerer" and bison in the cave at Lascaux, in southwestern France.

PLATE 11

The anonymous artists of the Middle Ages who created the stained glass windows of the great Gothic cathedrals literally painted with light. Their work has never been surpassed.

PLATE 12

A Tiffany lampshade, created by the son of the famous jewel merchant; perhaps the highest stained glass work since the Middle Ages.

PLATE 13

Dye powders can be made in practically any color imaginable, thanks to modern chemistry. The powders are dissolved in a solvent, such as water, when used for coloring fabrics or paper.

PLATE 14

The Death of the Virgin, *by Pieter Brueghel the Elder (1525–1569). By the High Renaissance, European painters had learned how to use perspective and to handle light in ways that produced utterly realistic, yet beautifully dramatic works of art.*
(Captions continue on page 175.)

PLATE 10

PLATE 11

PLATE 12

PLATE 13

PLATE 14

PLATE 15

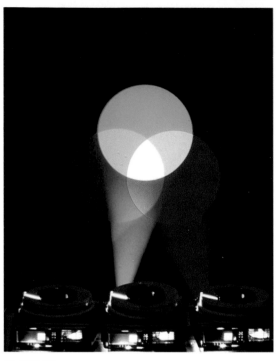

PLATE 16

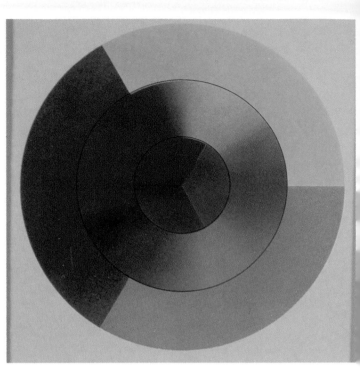

PLATE 17

PLATE 18

PLATE 15
Masaccio's Holy Trinity, *painted about 1427, was one of the earliest frescoes to make full use of perspective.*

PLATE 16
Additive color mixing. Modern technology reproduces experiments first done by Isaac Newton two centuries ago. Three slide projectors show how additive color mixing can produce a variety of hues from the three primary colors of red, green, and blue.

PLATE 17
A modern color wheel, showing the principles of additive color mixing.

PLATE 18
Subtractive color mixing. Colored filters can be overlapped to produce various hues. Subtractive color mixing is used in theater lighting, painting, and printing processes.

perceive are mixtures of blue, green, and red, is supported by the physiology of the retina, with its cone cells that are sensitive to blue, green, and red.

But does that tell the entire story?

In 1878 Ewald Herring asked an intriguing question: Why are there certain mixtures of colors that we do *not* see? We never see a yellow-blue mixture, for example. Mix yellow and blue and you get either unsaturated yellow, unsaturated blue, or blah gray. (Remember, gray is what completely color-blind persons see.)

Herring claimed that there are actually four basic colors: blue, green, red—and yellow. How can only three color receptors in the eye deal with four pure colors? Herring tried to explain it by suggesting that there are actually four sets of color receptors in the eye and that they work in pairs that are always opposed to one another so that we perceive the colors as we do.

But the physiological evidence showed only three color receptors in the eye. However, modern variations of Herring's ideas propose that the neural wiring between the retina and the visual cortex is such that the nerve impulses do act in opposition to one another. Thus, if the green and red receptors send signals of equal strength into the visual cortex, the cortex somehow cancels them out.

In the 1950s and subsequently, a new theory of color vision was proposed by Edwin Land, the inventor of the Polaroid camera. Land pointed out that the colors we perceive appear to be fairly constant despite the spectral power distribution of the light illuminating them. For example, a bed sheet appears white when it is hanging on a clothesline in the brilliant glare of the noonday Sun; it still appears white when it is softly illuminated by a single bedside lamp. Even skin tones illuminated by bluish fluorescent light are still perceived as skin tones; ghastly as they may seem, they are still recognizable.

Land suggested that this *color constancy* is an important clue to the fundamental mechanism of human color vision. He invented the *retinex* theory, the word coined from blending the words *retina* and *cortex*. As you might suspect, his theory involves the interaction between the retina and visual cortex.

The retinex theory is fairly complex. Basically it claims that the neural signals sent to the visual cortex by the retinal cells depend on the illumination of each region of the retina. The signals fired off by each nerve cell of the retina are compared against one another on the basis of the light's reflectance curve, and not its SPD. According to Land, then, the colors we perceive depend more on the way surfaces reflect light than on the light illuminating those surfaces.

Suffice it to say that the details of human color perception are still largely unknown. What is known is that the retina possesses three types of color receptors, sensitive to blue, green, and red. The nerve impulses sent from the retina to the visual cortex are extremely complex. Lateral inhibition, contrast effects, and color constancy all appear to play a role in the way colors are perceived.

Artists have unconsciously utilized the way we perceive light and color to produce their works. Unconsciously, at least, until modern times when the scientific understandings of human vision have allowed artists to approach their work more knowledgeably.

Great artists, even those who came long before our modern understanding of human vision, used their own eyes and minds, their own observations of the world around them and of human behavior, to produce works of immortal beauty. Such artists *think* in color, as Vincent van Gogh did when he wrote to his brother Theo about his painting of his bedroom in Arles (see Plate 19 following page 238):

> The walls are pale violet. The ground is of red tiles. The wood of the bed and chairs is the yellow of fresh butter, the sheets and pillows very light greenish lemon. The coverlet scarlet. The window green. The toilet-table orange, the basin blue. The doors lilac. . . . The broad lines of the furniture . . . must express absolute rest.

Scientists can explain the finished result of such thinking, to a degree. No one can explain *exactly* how the mind perceives those wonderful colors or the emotional impact that they make on the viewer.

Yet, even while van Gogh was living in Arles, other Frenchmen were inventing a new way to capture images of the world around us. The scientific understandings of optics and chemistry led to photography.

13

Through a Glass . . .

ANY BOY or Girl Scout knows that you can ignite paper by focusing the rays of sunlight with a magnifying glass. And adults, as well as children, are fascinated by the reflected world they see in the mirror.

Light can be manipulated. It can be bent, reflected, even sliced into its component colors by lenses, mirrors, and prisms. This has allowed clever humans to construct tools that use light for purposes of seeing the invisibly small or viewing the impossibly distant, tools that improve our failing eyesight and decorate our homes.

Mirrors, lenses, and prisms. Pieces of glass that do strange and wonderful tricks with light.

Glass itself is a rather strange and wonderful material. As solid as stone, yet it is transparent as air. Glass allows light to pass through it virtually unhindered. It is literally a liquid but not a fluid. That is, the molecular structure of glass is like that of a liquid (water, for example), where the molecules are not rigidly aligned with one another in a crystal lattice structure. In normal solids, such as a metal window frame, the molecules are tightly organized, like a team of acrobats locking arms and standing on one another's shoulders to form a human pyramid. In glass, the molecules resemble more closely a gathering of people relaxing at a cocktail party.

Although at the molecular level glass is really a liquid, for all practical purposes the stuff behaves like a solid. It does not flow like a liquid at ordinary room temperature; scientists say it has the viscosity of a solid. If you bump your nose on it, it will hurt just as much as if it actually were a solid.

The glassmaker at work. Physically, glass is a "liquid" that solidifies at normal room temperature.

Glass is created by heating silica, the stuff that makes up sand, until it liquefies. Certain additives are mixed with the silica to provide strength and, when desired, color. Then the glowing-hot liquid is rapidly cooled. It turns hard, although it does not crystallize. The "liquid" solidifies into a substance that allows light to pass through it.

Glass occurs naturally when hot magma from the Earth's molten interior spews out onto the surface and is cooled so rapidly that the silica cannot crystallize. Obsidian is the result: natural glass. Some volcanic rocks are partially glassy.

Who invented the first lens, we do not know. It happened too far back in history for any firm record to remain. Mirrors, also, were common by the time history began to be recorded. Ancient Egyptian and Sumerian women used makeup and examined the results in mirrors of polished metal. Even earlier, legends arose from the fact that you can see yourself reflected in a pool of water.

In Greek myth, for example, Narcissus was promised a long life by the gods—providing he did not look upon his own features. But once he saw his face reflected in a pool of water, he rejected his lover and spent his days staring at himself. Until he fell into the pool and drowned. Thus the flaw of narcissism, an unhealthy fascination with one's self, especially one's physical image. It is a flaw that is especially prevalent in Hollywood. And Washington.

Archimedes knew enough about mirrors to focus sunlight as a weapon of war against the Roman fleet that invaded Sicily in 212 B.C. Eyeglasses were invented during the Middle Ages; by the thirteenth century they were common enough both in Europe and China. In 1608 Hans Lippershey invented the telescope, and the following year Galileo used it for astronomical observations. In 1665 Newton began the experiments and theorizing that led to the publication of *Opticks* in 1704. Benjamin Franklin invented bifocal spectacles in 1784.

All of these discoveries and inventions hinge on one vital fact: Light travels at different speeds in different media. In the vacuum of space, light speed is roughly 186,000 miles per second. It is slower in air and slower still in glass. There are many substances that are opaque to light: a brick wall, for example. Other substances, such as water and glass, will transmit light. Still others reflect light.

Polished metals reflect light nicely, and behind every glass mirror is a thin layer of silvering, actually made not of silver but of the metal aluminum. It is the silvering that does the reflecting. The glass on top of it is there to protect the thin aluminum layer and hold it together.

Mirrors can be eerie. Sometimes you get the feeling that you are looking into a whole other world that exists on the other side of the glass. The image in the mirror has all the depth and color and reality of the world on *this* side of it. Physicists call the image seen in the mirror a *virtual image.* To our perceptions, it

A mirror can give the impression that an entire other world exists on the other side of the glass, exactly similar to our own world, only reversed left for right.

seems as if the mirror is actually a window showing a view that is almost—but not exactly—the same as the world in which we are standing.

Newton and his fellow scientists showed how rays of light are reflected from a mirror and why the scene we observe is reversed, right for left. It is a simple matter of geometry, once it is explained.

It works this way: Suppose you are standing before a mirror. Behind you is a chair. For this thought experiment we will ignore everything else in the room and concentrate solely on that chair. Light (from the Sun or a lamp) is reflected off the chair. If you were turned to face the chair you would see it directly by the light it reflects. Its color would be determined by which wavelengths of light it reflects best and which it absorbs.

But you are not facing the chair; you are facing the mirror. Light bounced off the chair strikes the mirror and is reflected to your eyes. Now comes the slightly tricky part. The rays of light your eyes receive have no "flag" on them that says they were reflected off the mirror. Your visual system merely detects rays of light that your brain decodes into an image of a chair. The rays of light seem to be coming from some distance *behind*

The light reflected by a mirror is perceived by the brain as coming from some distance behind the mirror's surface. This is called a virtual image.

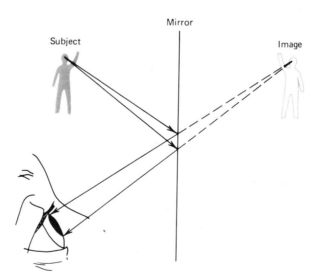

the mirror—the same distance, in fact, that the chair is from the mirror's surface.

The human visual system has no way of deciphering whether or not the light it is receiving has been reflected off a mirror. It sees a chair, and it tells the conscious part of your brain that the chair is in front of you at a certain distance. It takes those little gray cells of the cerebral cortex to determine that you are looking at a reflection, a virtual image, and not the chair itself.

In the first century A.D. the Greek engineer Hero laid down the basic rules by which mirrors reflect light. Hero was one of the few ancients who interested himself in things mechanical. Living in Alexandria, he "invented" not only the steam engine but jet propulsion as well! Hero built a hollow metal sphere with two small pipes extending from it. When the sphere was filled with water and the water heated until it boiled, steam escaping from the pipes made the sphere spin.

But the Greeks did nothing with this "toy." It was not until Newton elucidated the principle of action and reaction, 16 centuries later, that the groundwork was laid for the eventual development of jet engines and rockets. And it took about the same length of time for steam to be harnessed for practical engines by James Watt and others.

However, Hero's experiments with mirrors were the first to bring some practical order to the phenomenon of reflection. He showed that the angle at which a ray of light strikes a surface is the same as the angle at which it is reflected by that surface. If the *incident* angle is, say, 30 degrees (measured from a line perpendicular to the mirror's surface), then the *reflected* angle will also be 30 degrees. It's like bouncing a rubber ball off a smooth wall: The angle at which it rebounds is equal to the angle at which it struck the wall.

The amount of light that a mirror reflects depends on the quality of its glass and its silvering. Some light is always absorbed, so that even the finest mirrors do not reflect 100 percent of the light striking them.

Lenses are different from mirrors. While even the clearest and purest glass will reflect some light, lenses are designed to allow light to pass through them. However, since light travels at different speeds in different media, lenses can bend light, focus it, or spread it out. Here we are dealing with the phenomenon of *refraction*, the bending of light rays, rather than reflection.

The most common example of refraction is the way a stick appears to bend when it is put partway in a pool of water. Light travels more slowly in water than it does in air, and thus the angle at which the stick is held seems to be different in the water than in the air.

The optics of lenses can be complicated, even though fascinating. Suffice it to say that lenses come in two basic types, with several variations within each type.

First, there is the *converging* lens. This is shaped so that its middle is thicker than its edges. Its surface is convex, or bowed outward. Converging lenses, as the name implied, concentrate light rays, make them converge to some focal point. Sherlock Holmes's magnifying glass is an example of a converging lens.

When a converging lens is used to start a fire, by focusing sunlight on paper or tinder, it is obvious that the lens not only bends light rays but also intensifies the energy of the light by concentrating it on a small spot. That spot where the light is concentrated is called the *focal point*.

The other basic type of lens is, as you might have guessed, the *diverging* lens. Its surfaces are concave, thicker at the edges than the middle. It bows inward. Diverging lenses spread light rays. They are no good at starting fires.

A lens that bows outward on both sides is called a *double convex* lens, while one that bows inward on both sides is a *double concave*. It is possible to grind a piece of glass so that it is convex or concave on only one side, while the other side remains flat. Such lenses are called *plano-convex* or *plano-concave*. Finally, it is possible to make a lens that is convex on one side and concave on the other. Such a lens is called a *meniscus*. Most eyeglasses have meniscus lenses.

Roger Bacon, the thirteenth-century English friar who, like Galileo, had his scientific works suppressed and was himself imprisoned by the Church, was an experimenter who tinkered with gunpowder as well as lenses. He showed that properly shaped lenses could help people with failing vision. Following this notion,

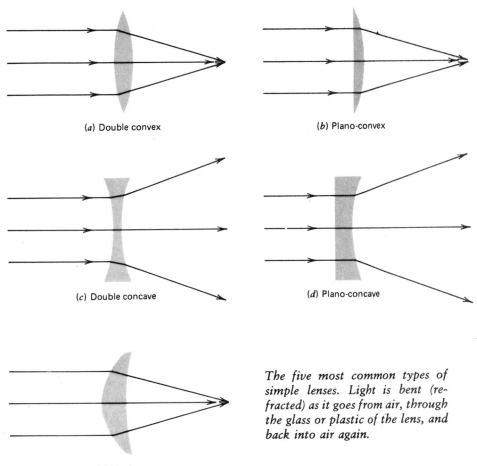

(a) Double convex

(b) Plano-convex

(c) Double concave

(d) Plano-concave

(e) Meniscus

The five most common types of simple lenses. Light is bent (refracted) as it goes from air, through the glass or plastic of the lens, and back into air again.

eyeglasses soon made their first appearance in Italy; their invention is variously attributed to Salvatore D'Armate or Alessandro di Spina of Florence. Spectacles may have already been in use in China. The evidence is unclear as to where the idea originated and whether it traveled East to West or vice versa.

As we saw earlier, the human eye focuses light to achieve a sharp, clear image. Much of the focusing is done by the cornea, the tough outer coating of the eye. The fine focusing, though, is performed by the lens that lies just behind the pupil. The lens is stretched or shortened by muscles attached to it, much as the skin of a drum is tightened or made looser by increasing or decreasing the tension on it.

With age, the lens begins to loose its ability to change its shape, and vision begins to suffer. The most common problem is farsightedness, where the lens can no longer focus on objects close to the eye. This affects one's ability to read, to sew, or to do any other kind of close work. The lens can no longer focus the incoming light on the retina. Instead, the focal point of the lens is well behind the retina. The result is a fuzzy, blurred image—like the picture seen on a screen when a slide projector is not focused properly.

A converging glass lens can correct farsightedness by focusing the incoming light enough to produce, together with the cornea and the eye's own internal lens, a sharp image on the retina.

Myopia, nearsightedness, results when the eye's lens can no longer focus on distant objects. The focal point of the lens falls short of the retina, and it is only when an object is quite close to the eye that it can focus adequately. A diverging glass lens can correct the situation and bring the focal point onto the retina for clear vision.

Most people tend to become farsighted with age. I wear bifocals to read; I am wearing them as I type these words. The bottom half of my eyeglasses are converging lenses so that I can see close-up figures. The top half is clear glass so that I can see distant objects without distortion. My daughter is rather myopic and needs glasses with diverging lenses to see things from afar. When she was a schoolchild the problem became apparent, so my wife and I took her to an optometrist who measured her vision and prescribed corrective lenses. Driving home after we had picked up her glasses, she exclaimed, "The town hall tower has a clock in it!" She had not been able to see that detail without the eyeglasses. Or so she claimed.

Eyeglasses have become stylish; they are no longer the square, wire-rimmed spectacles of Franklin's day. Still, some people prefer not to wear glasses; they use contact lenses. The optical principles of contact lenses are the same as those of spectacles; the lenses are merely small and light enough to be slipped directly over the corneas. They can even be colored to suit the whim of the wearer.

Until recently, there was little that medicine could do to correct problems of the cornea. Old age, disease, or genetic disposition can produce cataracts, in which the cornea hardens and turns opaque—sometimes a horrifying milky white. The rest of the eye may be perfectly healthy: pupil, lens, and retina in fine working order. But if the cornea clouds over, it is the same as if a curtain is drawn across a window. The result is blindness.

In modern times, surgeons have been able to cut out the cornea. More recently, the energetic beam of laser light has been used instead of a scalpel; a fitting irony, using light to restore sight.

However, with the cornea gone the eye has lost a good deal of its focusing power. Artificial lenses are now implanted over the eye to replace the cornea, but it is still necessary to wear eyeglasses for close vision, glasses with thick lenses that make the wearer appear weirdly owl-eyed. In 1987, though, the U.S. Food and Drug Administration gave its approval for clinical trials of a new bifocal lens that permits both near and far vision without the need for eyeglasses. The plastic lenses were successfully implanted in two patients at Ohio State University. Similar lenses have been undergoing clinical trials in Britain.

Microscopes and telescopes are optical devices that use two or more lenses to allow us to see objects that are beyond the range of unaided human vision.

We have already seen that the telescope was invented around 1608 by the Dutch spectacle maker Hans Lippershey. The microscope was invented slightly earlier, in 1590 by Zacharias Jenssen, also in Holland. The Dutch were at the forefront of optics then, and a considerable industry in lens grinding existed in Holland. Like the telescope, the microscope was significantly improved by—who else?—Galileo.

While microscopes that employed a single magnifying lens had been in use since the fifteenth century, Jenssen produced the first compound microscope, using two lenses. The convex *objective lens,* at the lower end of the microscope tube, magnifies the object to be viewed. Then the second lens of the *eyepiece* magnifies that image even further. This allowed much greater magnification of the objects being viewed. If the objective lens has a magnification power of 100, for example, and the eyepiece a power of 10, then the specimen being observed is magnified by 1,000 times. Modern optical microscopes typically offer magnification powers

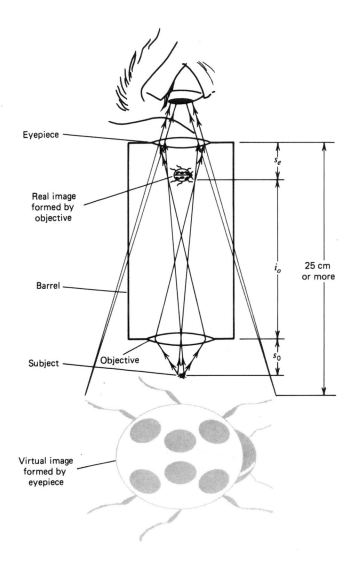

Eyepiece

Real image
formed by
objective

Barrel

Subject

Objective

s_e

i_o

25 cm
or more

s_0

Virtual image
formed by
eyepiece

*In the compound microscope, lenses
produce a highly magnified image
of the subject being viewed.*

of about 1,000 to 2,000. Electronic "boosters" added to optical microscopes enhance their powers even further. (The details of electro-optical boosters are discussed in Chapter 19.)

It is impossible to contemplate the growth of biology or medicine without the microscope. This piece of optical equipment allowed scientists to peer into the workings of the cells, the basic building blocks of living organisms. By 1658 red blood cells were discovered by Dutch biologist Jan Swammerdam. By the 1670s Antony Van Leeuwenhoek became the first person to observe single-celled creatures such as bacteria and protozoa. In 1831 the Scottish botanist Robert Brown (for whom Brownian motion is named) discovered that cells contain nuclei—the first step on the road that eventually led to unraveling the double helix of DNA and its genetic code. The work of Pasteur, the entire germ theory of disease, the giant strides forward in medicine that have raised our life expectancies and promoted our good health, would have been impossible without the microscope.

Its descendants, the electron and ion field microscopes, which use beams of subatomic particles instead of light, have extended our vision down to the very atoms that make up all matter.

Before these discoveries could be made, the original optical microscopes had to be greatly improved. One of the problems with microscopes—and telescopes, too—was that their lenses produced images that were often tinged with annoying rings of color, usually red and blue. This is called *chromatic aberration,* meaning essentially "colored distortion." The lenses in these earliest optical instruments refracted (bent) the different components of white light at slightly different angles; they acted as prisms as well as magnifying lenses, spreading the white light and distorting the image under study. The more powerful the lenses, the greater the distortion. This kept the optical microscope from being developed to its full potential.

It was not until the 1880s that the problem was fully solved by the use of *achromatic* lenses, where two lenses made of different types of glass, each with a different index of refraction, are cemented together. Where one part of the lens spreads the incoming white light, the other part concentrates it. One half of the lens counteracts the aberration caused by the other half, and the resulting image is free of chromatic aberration.

Telescopes were also plagued by chromatic aberration, but there was a different way around the problem that Isaac Newton invented. It is called, appropriately, the *Newtonian reflector* telescope.

The earliest telescopes were refractors with compound lenses: that is, a convex lens at the far end of the tube and a concave lens at the eyepiece. Since Galileo popularized the telescope's use with his pioneering discoveries, this type of refractor is to this day

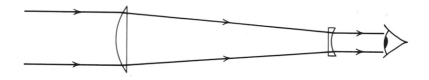

The simple Galilean telescope was limited in the magnification it could produce and was plagued by annoying rings of color around the image, called chromatic aberration.

called a Galilean telescope. Another great astronomer, Johannes Kepler, who was plagued with poor eyesight, suggested placing a convex lens at the eyepiece to increase the telescope's magnification power. While this produced an inverted image, astronomers did not mind seeing the heavens upside down. Inverted astronomical pictures can confuse the unwary, however.

Newton's work laid a firm foundation for the understanding of optics, which eventually allowed telescopes to become better and bigger. In his time (and for some while afterward) chromatic aberration forestalled all attempts to make truly large refractors. So Newton hit upon the idea of using a mirror to "collect" light and reflect it to an eyepiece. The mirror could be made as large as contemporary technology could produce, without chromatic aberration, since the mirror merely reflects the light and does not bend it through a lens.

Trying to get around this limitation of refractors, Newton thus invented the reflecting telescope. In the Newtonian reflector a primary mirror gathers in the light and reflects it to a secondary mirror, which in turn reflects the light into a magnifying lens that serves as the eyepiece.

The great advantage of the reflector is that it can be made *big*. All the largest astronomical telescopes of the world are reflectors,

Isaac Newton hit upon the idea of using a mirror to "collect" light; his reflecting telescope got around the problem of chromatic aberration and allowed much larger telescopes to be built.

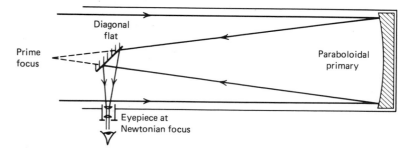

from the Hale Telescope atop Mount Palomar to the Hubble Space Telescope designed for use in orbit. The primary mirror for the Mount Palomar device is 200 inches in diameter; the Space Telescope's primary mirror is 94 inches across.

Telescopes dramatically changed the field of astronomy, of course. As astronomers and optics technicians learned to make mirrors larger and more precise, with surfaces polished down to within a few microns' smoothness, telescopic observers were able to peer deeper and deeper into the sky and to make new discoveries.

William Herschel was a musician who fled his native Hanover, in Germany, and eventually became court astronomer to Britain's George III (himself a Hanoverian). Herschel built the largest telescopes of his day, capped by a 48-inch reflector. In 1781 he discovered the planet Uranus, the first planet not known to the naked-eye astronomers. This led to searches for other members of the Solar System and to the discoveries of Neptune (1846) and

Kitt Peak National Observatory, near Tucson, Arizona. The largest astronomical telescopes are all reflectors.

Pluto (1930). The search goes on; some astronomers are convinced that there are other planets to be found, orbiting the Sun even farther off than distant Pluto.

The Herschels were a remarkable family. William's sister Caroline not only worked as his assistant, but she discovered eight comets. His son, John, made important contributions to astronomy, chemistry, physics, and mathematics.

John Herschel also coined the word *photography* in 1839.

For centuries it had been known that salts that contain silver in them would turn black when exposed to light. In 1802, Thomas Wedgewood and English chemist Humphry Davy studied a method of reproducing drawings and silhouettes on paper or leather that had been treated with silver chloride or silver nitrate. The images would fade in time, however.

Herschel showed in 1819 that the images could be permanently fixed if they were treated with certain chemicals that contain sulfur, called hyposulfites.

That was a crucial step: the ability to make a permanent image from the light-darkened silver salts. The camera obscura had been known since before Alhazen's time, nearly a thousand years earlier. Joseph Nicephore Niepce was a French lithographer; he etched drawings onto metal plates, which were then reproduced in a printing press. In 1826 Niepce used a small camera obscura to make an eight-hour exposure on a sheet of pewter that had been coated with an asphalt solution. Although the results were hardly startling, Niepce had brought together two powerful ideas: the pinhole camera (which is what the camera obscura was) and the possibility of producing an image from a chemical reaction triggered by light.

Niepce began to work with Louis Daguerre, a Parisian painter and scenery designer for the Paris stage. Daguerre created dioramas, sweeping backdrops in which he tried to blend real objects on the stage with painted backgrounds to give the audience a more realistic effect.

In 1829 Daguerre began to work with Niepce on the possibility of producing photographic images. Although the word *photography* had not yet been invented, the process was *almost* there. Niepce's technique required hours of exposure. Daguerre experimented with copper plates treated with silver salts as the "film."

It took years of experimentation. Niepce died in 1833, but Daguerre pressed on. One of the critical advances he made was to replace the pinhole of the camera with a lens that could admit more light, thereby making the pictures brighter, yet still focus the light properly on the plate. His efforts eventually succeeded. Daguerreotypes were the first practical photographs, and a new way of visualizing the world had been created.

Photography quickly found uses in journalism, in art, and in science. Samuel F. B. Morse, inventor of the telegraph, was the first American to try to mate the camera with the telescope for astronomical observations. Photographs recorded the everyday world and allowed the general public to see faces and scenes from virtually anywhere in the world. Mathew Brady's daguerreotypes of the American Civil War were among the first photographs to show the face of battle as it really is. Today photography brings us images of war and beauty, famine and high fashion, art and nature, every day.

However, the camera would have remained a cumbersome device to be used only by experts if not for Englishman Fox Talbot and American George Eastman.

Until 1840, photographs were made entirely inside the camera. The lens was focused on the subject, the light fell on the metal

A Daguerreotype camera. While such cameras allowed photography to flourish in the hands of professionals, it was not until George Eastman invented compact roll film that photography came into the hands of everyone.

plate that served as the film, and the photographer waited until the image on the plate was bright enough to be seen. It could take hours.

Talbot invented real film. He used paper treated with silver chloride crystals. This produced a negative image: The areas where light hit the silver chloride turned dark; unexposed areas remained light. After fixing the image to make it permanent, the negative could be pressed against a similarly treated paper and exposed to sunlight to make a positive print. From a single negative Talbot could make a virtually unlimited number of contact prints.

In 1840 he discovered that he did not have to wait until a visible image built up on the film. Even though no image could be seen on the treated paper, Talbot found that if the paper were treated chemically, the *latent image* would appear. This process, now called "developing" the film, meant that much shorter exposure times could be used. There was no need for the camera shutter to be open for hours or even minutes. A much briefer exposure would put a latent image on the film, which could then be developed into a negative that could be used for printing positive reproductions.

Talbot's paper film could not compete, however, with the new chemical emulsions that were used on glass plates. Claude Niepce, a cousin of Joseph Nicephore Niepce, invented the glass-plate technique in 1847. The emulsion-on-plate system gave far better picture quality, and paper film was almost forgotten.

But in 1884 Eastman produced a celluloid film that not only produced picture quality that competed with the plates but also could be rolled into a compact cylinder. Once exposed, the celluloid film could be developed and printed onto paper. In 1888 he came out with a hand-held camera that was so simple that anyone could use it. He coined the word Kodak as the name of his revolutionary camera.

Photography came to the hands of everyone, and Eastman created a business empire based in Rochester, New York, that still dominates much of the world of cameras and film.

Color photography began with no less a man than James Clerk Maxwell. In 1861 he demonstrated the first color photographs, which he made by exposing the same photographic plate three times through filters of red, green, and blue.

Scotsman that he was, Maxwell chose as the subject for his

color photograph a tartan plaid ribbon, bright with many colors in an intricate Highland clan pattern. Through his studies of light and color, Maxwell realized that exposing the film through filters sensitive to the three primary colors should produce a photograph that showed all the colors of the original plaid.

He was right, but for the wrong reason. Years after Maxwell's successful demonstration, it was found that the photographic emulsion he used was insensitive to red light. How could it have reproduced full color when it was only "seeing" blue and green? It was not until 1961 when scientists at the Eastman Kodak Research Laboratories in Rochester discovered that the red areas of the plaid reflected ultraviolet light as well as visible red, and it just happened that the supposedly red-sensitive chemical in Maxwell's emulsion was actually sensitive to UV. The ultraviolet light turned the emulsion red exactly where red light would have. A remarkable coincidence that confirms Joe Garagiola's dictum from the sport of baseball: "The great hitters get all the breaks."

Motion pictures began with a bet.

For uncounted generations children (and adults) had been amused with toys that utilized the persistence of vision to present the appearance of a moving picture. The Zoetrope was typical of such toys: a revolving drum with pictures inside it and slits to look through. Spin the drum and what were individual pictures of, for example, a running horse blended together into a moving picture of a horse galloping along.

In 1872 Leland Stanford, who was then governor of California, hired Eadweard Muybridge to help him settle a bet. Stanford had wagered $25,000 with a friend that when a horse gallops, at some point in its stride all four hoofs are off the ground. Muybridge set up 12 still cameras along the edge of a racetrack and snapped 12 consecutive photos of a horse galloping by. The pictures showed conclusively that all four hoofs did indeed leave the ground in midstride.

Muybridge's technique of taking sequential photos combined with the phenomenon of persistence of vision provided the basis for motion pictures. Eastman's celluloid film, which could be rolled into a compact package, made it practical.

Thomas Edison is credited with inventing the motion picture camera, the 1888 Kinetograph, and the motion picture projector,

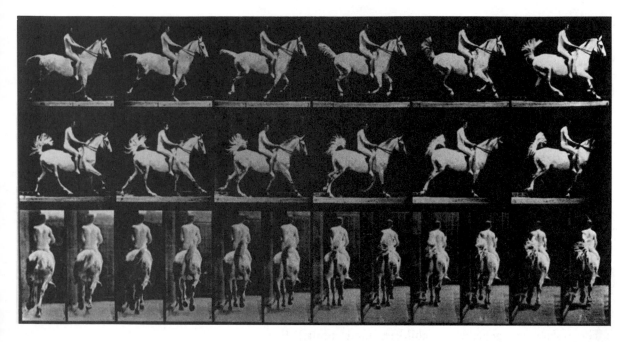

Motion picture photography began when Leland Stanford, then governor of California, bet $25,000 that a horse lifts all four hoofs off the ground simultaneously when galloping.

the Vitascope, patented in 1896. Other inventors in other nations also claimed priority in inventing motion pictures, largely because Edison's first efforts were not projected on screens in theaters but were peep shows where only one person at a time could see the moving picture.

Edison built a motion picture studio and in 1893 began producing films on its indoor set.

Meanwhile, in France, the brothers Auguste and Louis Lumière* developed camera-projector equipment light enough to be carried anywhere and began producing the first movies outside of a studio, on location.

Interestingly, Edison's productions were all fictional, theatrical. The camera was used as if it were a single member of the audience watching a staged production. The Lumières, on the other hand, went outdoors and filmed natural events: the earliest documentaries.

*Another etymological delight. *Lumière,* in French, means "light," "lamp," or "enlightenment."

From these beginnings rose the motion picture industry.

The next great breakthrough in photography came in 1947 when the American Edwin Land invented the Polaroid instant camera. Land's big contribution was to invent a film that developed a positive image within seconds of exposure to light. Both the negative film and positive paper are sandwiched together in the film pack. When the film is pulled out of the camera, it passes through rollers that break a chemical pod, releasing the chemicals that develop the film and transfer its image to the positive paper.

Polaroid cameras have made instant photography easy not only for tourists but also for scientists who want to record transient phenomena and see the results at once.

From its humble beginnings, photography has become a mam-

Two of the giants: George Eastman (left), inventor of roll film, and Thomas Edison, who is credited with inventing the motion picture camera (among many other things).

moth worldwide industry, an art form, a tool for scientists of all types, a field of journalism, and a means for the average person to achieve self-expression and to record everyday life.

Photography had an enormous impact on painting. For centuries painters had striven to reproduce what the eye sees. Now a click of a shutter allows even the rankest amateur to make more faithful pictures than any painter could hope to create. Realism was stolen away from the painters. They moved toward other domains, other approaches to art, and away from realism.

Claude Monet and the other Impressionist painters would probably never have produced the masterpieces they created if the camera had not taken away from them the domain of photographic realism. Cezanne, van Gogh, Picasso, and all the later abstract painters have had to find their expression in a world where the camera is always "competing" with them. (See Plate 20 following page 238.)

Yet photography itself became an art form. Not only realism but also abstract images can be recorded on film. And once Thomas Edison perfected the motion picture camera and projector, movies could aspire to high art—and low commercial exploitation.

Cameras have replaced human vision at the eyepieces of astronomical telescopes and the biologist's microscope. Satellites carry cameras that gaze starward and back at Earth to photograph weather patterns, pollution plumes, and military construction sites. Spies use miniature cameras. Tourists carry megatons of camera equipment around the world every day; in many cases, a tourist's camera equipment is the most expensive and precious possession taken on the trip.

We photograph birth, baby's first steps, and all the rites of graduation, marriage, birthdays. Robert Capa caught the instant of death when he snapped an unforgettable photograph of a soldier as he was hit by a fatal bullet in battle during the Spanish civil war.

Photography has produced its own artists, such as Alfred Stieglitz, Edward Steichen, and Henri Cartier-Bresson.

One frosty afternoon in Massachusetts, years ago, I had taken my young children to a nearby beach where we could fly a kite without danger of it being snared by trees or overhead power lines.

It was quite chilly, and the long curving beach was deserted—except for one man in a windbreaker and a broad-brimmed hat, bending over a camera set up on a tripod.

As the children flew their kite I watched the photographer at work. He was a friendly older man, and we soon fell to chatting about the beauties of the area, even in winter. Almost as an afterthought, he introduced himself: Ansel Adams.

Eventually I saw the photos that he had taken that day. In them I saw that lonely beach and cloud-swept sky in ways I had never seen them before.

There was an artist who "did your looking for you."

14

Lights, Action . . .

ON THE FINE summer evening of Tuesday, August 4, 1914, Sir Edward Grey looked out of the window of his London home as the street lights began to shine against the encroaching darkness.

"The lamps are going out all over Europe," he said with infinite sadness. "We shall not see them lit again in our lifetime."

Grey was Britain's foreign secretary. By midnight, Britain and Germany were at war. World War I had begun, and Grey's melancholy prediction was not far from wrong.

It was a poignantly apt turn of phrase. Since the Ice Age we humans have associated light—particularly lights that brighten up the hours of night—with safety, comfort, and joy.

The taming of fire marked a turning point in human history. The gift of Prometheus changed humankind from freezing, frightened leopard bait into masters of the world and challengers of the gods. To this day, almost nothing is as fascinating to us as watching the dancing flames of an open fire. A campfire is not only necessary, it is comforting, a place for storytelling and singing songs. A home's hearth is the traditional place for hospitality, the symbol of safety and care. And what is more romantic than a lovely little blaze in the fireplace on a wintry evening, as the old song says, "while the wind on high sings a lullaby?"

We are diurnal creatures. We are active by day and sleep away the night. But the history of the human race is a history of going beyond early limitations, of expanding our habitats and our way of life, of breaking the bounds set by our physical form. We cannot run like a deer, but we can outspeed any animal. We have no wings, yet we can fly higher and farther than any bird. We need air to breathe, yet we are expanding our ecological niche into the vacuum of outer space.

The very first expansion of our way of life, the original and most important breakthrough beyond our physical limitations, was the invention of artificial light, which began with the taming of fire. Not only did fire make us safe from the nocturnal predators that hunted our ancestors. Artificial light allowed them to be active when other diurnal creatures could not be. It stretched our "daylight" hours until now we can stay up around the clock if we want to, working or partying or even reading a book that we just can't put down.

At Lascaux, Altamira, and other sites of cave paintings there is evidence that the artists worked by torchlight. The chambers in which the cave art appears had to be lit artificially; sunlight never touched them. Which came first, the lighting or the art? It seems very likely that Cro-Magnon people were already illuminating their caves with torches before the first artists among them started drawing.

Torches and lamps that burned animal fats or vegetable oils were the major means of illumination throughout prehistory and well into the modern era. Candles were a step forward, a means of burning animal fats slowly enough to provide good light for a reasonably long time. Cyrano de Bergerac, the swashbuckling hero of Edmond Rostand's immortal play (and a real person of seventeenth-century Paris), reports:

> Sunday . . . The Queen gave a grand ball, at which they burned seven hundred and sixty-three wax candles.

Science and the so-called Age of Reason brought about improvements in artificial lighting. Physicists and chemists began to understand how combustion works, allowing the design of more efficient lamps. By 1784 the Argand oil lamp enclosed the wick with a glass chimney that produced a controlled current of air to feed the flame with more oxygen than simply allowing it to burn the way a candle does. The Argand lamp thus provided a brighter, whiter, and cleaner light.

In those days, the best oil came from whales. The "iron men in wooden ships" who sailed from Nantucket and New Bedford (and many other ports in other nations) killed whales with hand-thrown harpoons and sold their oil to light the lamps of the world. In one of history's continuing series of ironies, it was the discovery of petroleum in Pennsylvania that saved the whales from extinction. Whalers were slaughtering the great cetaceans in every ocean, with no thoughts about the ecological consequences, when cheap plentiful Pennsylvania crude knocked the bottom out of the whale-oil market.

Long before the automobile came into being, petroleum became an important fuel for artificial lights—and for heating.

Right behind the discovery of cheap petroleum came the discovery of coal gas. Coal had been used as a fuel since the beginning of the Industrial Revolution in the late eighteenth century. Now coal producers began cooking coal gas out of their coal and piping it to customers' lighting fixtures. Nowadays, with the price of petroleum held hostage to the politics of the Middle East, technologists point out that the United States still has hugely abundant reserves of coal within our national borders. This coal can be converted into liquid fuel to augment or even totally replace petroleum. Far from being a daring new technological scheme, coal gasification is more than a century old. Germany ran its autos and trucks on coal-derived fuels in World War I—and again in World War II.

Coal gas lighting was restricted to sizable cities because the gas had to be generated out of coal at special refineries. It made no economic sense to build such costly facilities in places where there were too few customers to support the expense. So gaslights were essentially urban comforts, while rural areas remained with the older oil lamps. Later, when natural gas came into use for heating and cooking, it was also used as the fuel for street lamps.

No fuel is without its hazards. Coal-burning pours megatonnages of carcinogenic soot and sulfur oxides into the atmosphere and leads to acid rain. Natural gas is cleaner, but in my childhood neighborhood of South Philadelphia, hardly a year went by without a major fire or explosion of the gas mains. It was the only way we got playgrounds—temporarily empty lots amid our row houses. All fuels produce carbon dioxide when they burn, and by increasing the CO_2 in the atmosphere, they enhance the atmospheric greenhouse effect, inexorably raising world temperatures and leading to climate changes that could eventually have disastrous effects.

Natural gas is cleaner and cheaper than coal gas. But its use in lighting was overshadowed (pardon the pun) by a totally different technology: electric lights.

From the first fires of *Homo erectus* until roughly 1879, artificial light came from flames. Torches, oil lamps, candles, gaslights—they all produced light by burning something. That presented certain hazards. I suspect that if the entire human race were hit by a form of selective racial amnesia, so that no one remembered there was ever such a thing as oil or gas lamps, and a bright young inventor came up with the idea of a lamp based on burning natural gas, the environmentalists and safety inspectors of our modern society would lobby furiously against it. Open flames are dangerous, an invitation to conflagrations. And the gas that fuels the flame is toxic; in the good old days "taking the gas pipe" was a euphemism for suicide.

Then along came Thomas Alva Edison, who said, "Let there be *electrical* light." And it was so.

Edison's contribution to the human race has been largely misunderstood. He is credited with thousands of inventions, from the phonograph to the motion picture projector. The Wizard of Menlo Park is often portrayed as a solitary figure, tinkering with one contraption or another until finally he shouts, "Eureka!"

Edison did not invent the electric light bulb. He did something far more important: He invented the electric lighting system. In fact, Edison's greatest contribution to the world was his invention of the *process* of invention. Far from being a lonely genius working in a shed or basement, Edison gathered around him whole teams of bright young men and organized the first purposeful research laboratory. There are still solitary inventors tinkering in their basements or backyards. But today new technological developments come out of giant research laboratories, staffed by hundreds of PhDs and other specialists. That is Edison's doing.

Electric lamps had been known decades before Edison tackled the problem. They were big and unwieldy clunkers, used for street lamps or very specialized purposes. You needed engineers

Thomas Alva Edison. Although posed here as the lonely genius, Edison actually created the first industrial research laboratory, where dozens of scientists and engineers worked together. He did not invent the electric lamp, but he did develop the total system needed to light the world with electricity.

to install them and a team of technicians to keep them running. They produced light by creating an electrical spark between two carbon electrodes; a miniature bolt of lightning, cracking and fizzing away as long as electricity—lots of electricity—was fed into the electrodes. Behind the spark was a well-designed reflector to make certain that the maximum amount of light was directed out of the lamp and toward the area needing illumination.

Such arc lamps are still in use today. Next time an automobile agency advertises a special sale and tells you to "look for the lights in the sky," go on down and take a look at the searchlights that send those intense beams probing into the heavens. Hollywood traditionally uses such searchlights at premiers of new movies. In World War II they were used to spot bombing planes at night, although radar soon became much more effective.

Arc-lamp light is rather kind to the human complexion, although candles are even better and much quieter for an intimate dinner. However, arc lamps are still used in some places for street lights and for lighting the exteriors of public buildings.

Many nations claim that the electric incandescent lamp (the kind you have in your home) was invented by one of their boys. Sir Joseph Swan of England actually did produce a carbon filament lamp in 1860. (Incidentally, he also invented the dry photographic plate some 11 years later, a great step forward in photography.)

But Edison went far beyond Swan and all the others. He was not interested in inventing a lamp by itself. He wanted to electrify the world—literally. He realized that the electric lamp meant little unless and until there were electric power-generating stations to produce electricity and a grid of transmission lines to distribute it. In modern parlance, Edison was a systems man; he integrated all the pieces needed to make electrical lighting practical and desirable.

Electric power generators had been known since the 1830s. Michael Faraday, the bootblack's son who became one of the

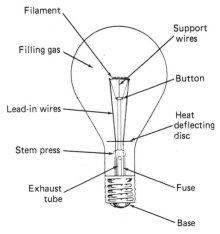

Filament

Support wires

Filling gas

Button

Lead-in wires

Heat deflecting disc

Stem press

Exhaust tube

Fuse

Base

The common incandescent light bulb, the basis for electric lighting all around the world.

finest scientists England has ever produced, built a little device back then that produced a current of electricity; he called it a *dynamo*. Legend has it that he gave a public lecture about his work, and after the lecture a matronly woman approached him with the question, "Your work with electricity seems very interesting, sir, but of what use is it?" Faraday is alleged to have replied, "Madame, of what use is a newborn baby?"

That response has become the motto of every researcher who is probing the frontier of knowledge. Yet there is another version to the story that may be even more appropriate. In this version, it is a member of Parliament who asks, "Of what use is it?" Faraday replies, "I don't know. But someday you will put a tax upon it."

Edison put Faraday's "useless" dynamo to work. The incandescent electric light changed the world. And sure enough, governments levy taxes on electrical power companies.

The incandescent lamp is essentially a glass bulb that has had the air vacuumed out of it. Inside the bulb is a filament of a material that will conduct electricity. When a current goes through the filament, the material of the filament is heated so that it glows. That glow is the light that we see coming from the bulb.

In 1879 Edison did indeed electrify the world by lighting up a 30-lamp system in Menlo Park, New Jersey, where he had set up his research laboratory. Three years later the Pearl Street station in downtown Manhattan gave New York the first urban electrical lighting system. The name Edison is still a part of many electrical power companies today, and justly so.

The earliest incandescent lamps used carbon-based filaments. Today the best filaments are made of tungsten, which has the highest melting point of all the practical materials available for such use. Tungsten lamps thus last longer than those using other filament materials. Although it had been known for years that tungsten would be the best material to use, no one knew how to draw this metal into thin wire until around 1900, when a process for doing such was invented—driven, no doubt, by the promise of a lucrative market in electric light bulbs.

Modern incandescent lamps are filled with an inactive gas such as nitrogen or argon, rather than vacuum. This enhances their longevity.

Edison's electric lighting system changed the world. Electric lights were far safer than open flames, although electric wiring presented some hazards of its own. However, once the power-generating equipment and transmission lines had been put in place, it became possible to use electricity for other purposes, in addition to artificial lighting. Electricity is a quiet and almost infinitely adaptable form of energy. Electric motors can run everything from the automatic garage-door opener to the Dustbuster hanging from its wall outlet to the electronic computer I use to write these words. Electric energy can bring the world's finest symphony orchestras to your home, courtesy of Mr. Edison's phonograph and its descendants. Radio and television transform electricity into sound and pictures. With electricity, we have access to more entertainment and education than the emperors of Rome could command.

Electric lighting also led to the popularity of evening news-papers. Most newspapers were distributed in the morning, when there was enough light to read their print easily. Reliable electric lights meant that evening newspapers, summarizing the events of the day *the same day that they happened,* swiftly gained popularity. The advent of electronic journalism has been a hard blow to evening newspapers; the day's events can be presented on radio or TV at the dinner hour. Although electronic journalism does not give the details that newspapers can, the electronic media can provide almost instantaneous coverage of news events.

We have become so dependent on cheap, abundant electrical power that when it fails, the modern householder suddenly loses lights, TV, stereo, the kitchen stove, even the home heating if it is run by an automated electrical thermostat. In regions such as New England, where hurricanes and blizzards can knock out the electrical power over entire states, many home owners have acquired auxiliary power generators.

Battery-supplied electricity can run portable radios and such. It also powers the boom boxes that youths perch next to their eardrums and play loudly enough to knock down chandeliers several blocks away. The only consolation to those who do not appreciate such noise is that the boom-box owner will undoubtedly be deaf by the time he's 30, and will no longer be able to assail the neighborhood with his version of music.

The incandescent lamp is the backbone of modern artificial

Paris, the "City of Light." Cheap, abundant electrical power not only provided lighting, but also brought on our modern era of electrical appliances and electronic entertainment, as well as the computer.

lighting. The other major kind of lighting is the fluorescent lamp, in which the bulb—usually a long, narrow cylinder—is filled with a noble gas such as neon or argon at low pressure. When electrical current is put through the gas, the atoms are stripped of some of their electrons in a process that produces photons—pure light energy. The color of the light is determined by the kind of gas in the tube. Each element gives off certain characteristic colors, and under the right conditions the colors can be bright indeed, even garish. The "neon" lights used in advertising are not all neon. (If you are interested in the how and why of all this, the physics behind fluorescent lamps will be discussed in the next chapter.)

Fluorescent lamps produce a maximum of light with a minimum of heat, since the process by which they yield light does not depend on heating a filament, as in the incandescent lamp. Fluorescents can also be made in virtually any shape, a boon to designers of advertisements, although often an eyesore to those of us who have to look at such garish sights.

Manufacturing light bulbs is today a major industry, and considerable research has gone into producing bulbs that are efficient and economical—and profitable. The manufacturer has to balance three essential factors. The first is brightness: How much light should the bulb put out for a given input of electricity? Second is color: No bulb will exactly duplicate the color of the Sun's light; what color should the light be? Third is lifetime: How long must a bulb last? It is possible to manufacture bulbs that will last virtually forever, but they will be dim and appear reddish to the eye.

Today's incandescent bulbs average about 1,000 hours of lifetime and produce a light that is noticeably yellower than sunlight, although whiter than candles or oil lamps. The glass envelopes of bulbs can be painted or otherwise treated to produce colored light or light that appears "softer" than the normal output of an incandescent bulb.

Electric lighting can literally turn night into day. In the theater, lighting effects do much more. Nowhere else can we see the dramatic effect that light and color have on our perceptions and our emotions as in the modern theater.

In the beginning, stage productions were lit by the Sun. The ancient Athenians would sit out in the open to watch Sophocles, Aeschylus, and their peers. Theaters were carved out of hillsides and sited so that the afternoon Sun lit the stage. The Romans stretched huge colored awnings over the seating area to soften the glare and heat of the sunlight.

As civilization moved northward, drama began to move indoors. Yet even in Shakespeare's time, the open-air Globe Theatre did brisk business during the summer months. The Blackfriars Theatre, completely enclosed, was used in winter. Candles were the main means of illumination, although torches and lamps were also used. As early as 1545 the Italian architect Serlio suggested placing flasks of colored water in front of the candles to produce lighting of different colors.

So far, the purpose of the lighting was merely to allow the audience to see the play. Usually the candles were attached to large hoop-shaped chandeliers that were hoisted up toward the ceiling on pulleys. Other candles or oil lamps were lined up along the foot of the stage and vertically in the wings, out of sight of

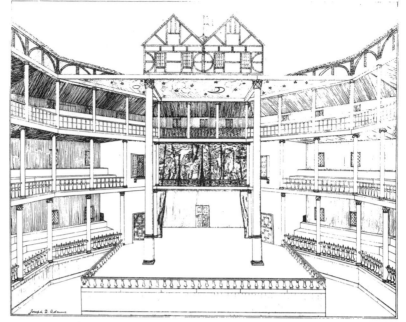

The theater before electric lighting.

the audience. Wall decorations of gold or other reflective materials helped to make the most of the available light.

David Garrick put footlights into the Drury Lane Theatre in London in 1765 and masked them from the audience's view with metal screens that also served as reflectors to enhance their light. The Argand oil lamp came into service after 1784, and soon afterward theatrical designers began using glass chimneys of different colors to change the lighting on stage. The chimnneys were raised or lowered by means of levers, and for the first time a member of the stage crew had the assignment of changing the lighting scheme on cue.

Since the lighting of the stage was neither very bright nor very subtle, the actors' makeup had to be strong enough for the audience to see clearly. The smell of greasepaint was very heavy indeed when layers of the stuff had to be smeared across an actor's face. We see a relic of those old days in the formalized masks used in ancient Greek dramas and in the traditional theaters of China and Japan. What appears excessive and exaggerated to us was originally designed so that customers in the farthest row of the audience could clearly and easily see who was smiling, who was frowning, who was the hero and who was the villain.

Even as late as twentieth-century burlesque, comics wore exaggerated makeup to identify their schtick. The immortal Bert Lahr began his career as a "Dutch comic" in those days when each ethnic group was lampooned mercilessly. A Dutch comic wore a heavy, squarish black greasepaint mustache; one look at him and the audience knew who he was supposed to be. Groucho Marx carried the tradition into the movies, where the greasepaint mustache and heavy-rimmed eyeglasses were certainly no longer necessary but provided an identifying trademark that subconsciously primed the audience for laughter.

Gas lighting was introduced in 1803 at the Lyceum Theatre, London; in 1816 the Chestnut Street Opera House in my native Philadelphia installed not only a coal gas lighting system but also its own gas-generating equipment, since the city was not yet "cooking (or lighting) with gas." Gaslights allowed much more flexibility. They could be brightened or dimmed with the turn of a knob. Soon theaters acquired elaborate control systems with pipes that carried the gas to individual lights and an intricate control board that allowed the lighting master to control the brightness of each lamp. This was, of course, the precursor to today's electrical control board.

While gaslights were brighter and cleaner than oil lamps and candles, they still posed the hazards of an open flame. In addition, they produced a lot of heat and an odor that was far from pleasant. But they could be made brighter or dimmer and even shut off entirely. For the first time, *darkness* was possible onstage!

Not only darkness came to the theater, but also the pinpoint of illumination that we have come to know as the spotlight. In 1816 an English engineer, Thomas Drummond, invented a lamp that produced light by burning a lump of chalky calcium oxide in the flame of an oxygen-hydrogen torch. The combustion of oxygen with hydrogen produces a very hot flame; the most efficient rocket boosters of our day use those two gases as their propellants. By narrowing the oxyhydrogen flame to a sharp point, a small area of the chalklike calcium oxide can be made to glow brilliantly with a white light that is soft rather than harsh.

Calcium oxide is, of course, lime. And Drummond's lamp, once it was introduced into the theater in the 1840s, soon became known as the limelight. Although it required a knowledgeable operator to run it because the block of lime had to be constantly turned slowly as the oxyhydrogen flame ate it away, the limelight became so useful to the theater that today the word is a part of our everyday vocabulary. To be in the limelight means that you are the star of the show, the one on whom all attention is focused.

The limelight, backed by a reflector that focused its light, made a wonderful spotlight. For the first time, an individual performer on the stage could be picked out in a spotlight, bathed in a bright yet eye-pleasingly soft glow that could follow the performer as he or she moved about the stage. The limelight could be screened by colored filters to produce a spot of virtually any hue desired. It seems so ordinary today that we barely think about it, but little more than a century ago it was a brilliant innovation.

Electric lighting, with all its flexibility of intensity and color, invaded the theater soon after Edison's first demonstration in 1879. To be sure, carbon arc lamps had been used earlier. The Paris Opera in 1846 used an electric arc lamp to simulate a shaft of sunlight falling on the stage, and by 1860 it had developed a system for producing rainbows and even simulated fountains on demand. The Opera company even produced the first spotlight, using a carbon arc lamp with a lens and a shutter to narrow the width of its beam.

But arc lamps needed electricity, and until Edison there were no electrical power stations available. The Paris Opera generated its own electricity. The other theaters of Paris—and everywhere else—stayed with gas lighting until Edison's breakthrough.

Electric lights replaced gas and carbon arc lamps. One vastly important consequence of that was that theaters became far safer. But more than that, the flexibility of electrical lighting allowed stage designers to begin to use light in ways that had never been dreamed of before. Light could create and enhance mood; light could be used like a painting to create a backdrop for the action on stage; light could even be used to alter the shape and size of the stage itself.

Adolphe Appia, a Swiss stage designer, was the first to grasp the new possibilities that electrical lighting offered the theater. Starting in 1891 he propounded his ideas of using sets and lighting to strengthen the dramatic action, of "painting with light" to create illusions and moods on stage. In books such as *The Staging of the Wagnerian Drama* (1895) and *The Living Work of Art* (1921), Appia insisted that the stage be turned into a fully three-dimensional setting rather than a mere platform in front of a flat painted backdrop. He realized that lighting could provoke emotional reactions from the audience, and he stressed the idea of light and color as a visual counterpart of music.

Appia's ideas were taken up by the English actor, director, and designer Gordon Craig, the son of the famous actress Ellen Terry and architect Edward William Godwin. Later, the American producer and playwright David Belasco, together with his electrical engineer Louis Hartman, brought a new sense of realism and power to the theater through lighting effects.

In his fine book *The Lighting Art,* Richard H. Palmer points out that the modern lighting designer aims to "light the production, not the stage." That is, lighting has progressed from merely providing enough light for the audience to see the stage to the point where it is just as important to the play (or opera, or ballet, or rock concert) as the sets, costumes, props, and dialogue. Lighting

Modern stage lighting can create mood, highlight action, focus the audience's attention on a particular area of the stage. Lighting directors literally "paint with light" to enhance the stage sets.

can evoke emotions ranging from the romantic glow of moonlight to the glaring danger of a raging fire to the deadly dread of deep, dark shadows.

He also claims that a play is much more than the lines spoken by the cast or even their actions on stage:

> In practice, the theater has rarely been dominated by the script. We tend to judge the theatrical activity of any period [of history] from the perspective of literary scholars, partly because they dominate our educational institutions, partly because the script is the most easily preserved part of a production. Yet theater today, as always, appeals largely to the eye, and there the lighting designer has final control.

As we have seen, lights of different colors and brightnesses can stir emotions within us. Usually bright light makes us feel more cheerful, dim light more apprehensive. Thus the lighting director's rule of thumb: bright light for comedy, more subdued light for serious drama. A gloomy mood can be conveyed, even created, by a low light level or by a flat, even light with no contrast; what can be gloomier than a dull gray sky, whether it is real or re-created in a theater? On the other hand, put a bit of contrast into a gloomy scene and it can become romantic; the dull gray sky is transformed into a moonlit evening.

In retail stores, warm-colored lights produce more sales than cool lights. In the theater, the decisions about color revolve around the emotions that various colors evoke in the audience. Violet light is considered the most melancholy color of all, since true violet is the coldest of all visible colors. Purple, a mixture of blue and red, is less heavy.

Blue light tends to make skin tones look dead, although green is regarded as the least flattering to the complexion (except for those rare colleens of true flame red hair and emerald green eyes). Yellow is perceived as bright and cheerful. Since yellow or amber light produces white when mixed with its complementary color, blue, these two colors are frequently used as the footlights and cross lights of the stage.

Orange flatters most skin tones and is the color associated with firelight and candles. Red is the most popular color among women and the most stimulating physiologically. It can also be used to suggest danger, anxiety, or aggression. And in sunsets.

In addition to mixing the right colors for the moods to be created and conveyed, the lighting designer must control eight "distinct properties of light," according to Palmer. They are:

1. *Intensity.* The amount of light actually reflected from the stage to the audience's eyes. This is controlled basically by the number, type, and wattage of the lights used. It is also influenced heavily by the scenery, costumes, and makeup of the actors.

2. *Color.* The hue, saturation, and brightness of the objects on stage. This is influenced by the color of the lights used, which in turn is a combination of the colors of the lights themselves, the color of filters used in conjunction with them, and the slight shifts of the lights' colors when they are dimmed.

3. *Direction.* Where is the light coming from? It might be afternoon sunlight streaming through a picture window or a single intense beam of light representing the wrath of God beating down on a solitary figure on stage.

4. *Form.* Light can create shapes, and shadows can insinuate other shapes or even an infinite expanse stretching off into nothingness. Pools of light can be arranged to be sequentially smaller and dimmer, suggesting distance, for example.

5. *Diffusion.* Light can be a hard pinpoint beam or a soft wash of color, depending on the effect desired. From a tightly focused spotlight to banks of lights shining through translucent filters, the diffusion of light affects the audience's perception of the color, form, and composition of objects on the stage.

6. *Frequency.* Lighting can be gradually dimmed or brightened to enhance a dramatic point or suggest the passage of time. Or it can be suddenly changed, as in the blackouts that often end comedy sketches. Strobe lighting, keyed to the eye's flicker rate, can generate intense emotional response at a rock concert.

7. *Motion.* A spotlight may follow an actor across the stage, or motion may be simulated by turning on a set of lights in sequence. As with frequency, motion is a matter of manipulating light over time.

8. *Luminousness.* This is the most subjective of all the qualities of light. It refers to the subtle interplay of brightness, intensity, color, and many other variables. Palmer says, "The liveliness of a laser image, the glowing quality of a fluorescing color, the flatness of a carbon arc light, the radiance of an incandescent source are all recognizable examples of this property. It is usually a product of the type of illuminant or filter color."

Given that a stage lighting designer can control those eight qualities, what is he or she trying to achieve?

First is the matter of selective visibility. Lighting controls what the audience can and cannot see. It can highlight certain areas of the stage or even spotlight individual performers. It can also hide things the audience is not supposed to see. It is important to know exactly what level of lighting is too dim for the audience to see; this produces darkness on stage, as far as the audience is concerned, although actors (or stagehands) can move through the shadows to prepare for the next scene.

Stage blackouts depend to some extent on how bright the light was immediately before the blackout and how long the blackout lasts. The eye adapts to low levels of light over time, and what seems impenetrably dark at one moment gradually becomes visible as the eye adjusts. Lighting designers usually try to stage blackouts after the stage has been brightly lit. They also like to make certain that any lights used during a blackout, such as hand lights the actors may require to find their proper positions on stage, are either far in the blue or far in the red end of the spectrum; these wavelengths are the most difficult for the audience to detect while the eye is adjusting to the blackout.

Lighting can help establish the time and place of the scene on stage. The brightness, color, and other factors of the lighting can tell the audience almost immediately if it is morning or evening, indoors or outdoors, heaven or hell that is being portrayed on the stage. Color can be crucially important, since colors have a strong subconscious effect on our emotions. The colors used in stage lighting can create or enhance the impact of a scene.

Proper lighting can also create three-dimensional shapes out of flat surfaces by suggesting to the audience's visual perception that the shadows and lights represent actual shapes and forms.

Finally, lighting can help to establish the style and rhythm of the production. The lighting schemes used for a lighthearted musical such as *Damn Yankees,* for example, are very different from those used for the much more serious *Man of La Mancha.*

Nowhere has lighting been used to such dramatic effect as at the 1934 Nazi party rally at Nuremberg. The architect Albert Speer, who later became Hitler's armaments minister, surrounded the huge Zeppelin Field with 130 antiaircraft searchlights. Their intense beams, every 40 feet around the perimeter of the vast stadium, stabbed straight up into the night some 20,000 feet and made the dark sky glow. Speer said:

The feeling was of a vast room, with the beams serving as mighty pillars of infinitely high outer walls. Now and then a cloud moved through this wreath of lights, bringing an element of surrealistic surprise to the mirage.

The British ambassador wrote, "The effect . . . was like being in a cathedral of ice."

The lessons learned from stage lighting were adapted for the motion picture studio in Hollywood's early days. Then film directors, lighting engineers, and camera operators went on to create their own lighting effects. Combining the needs of the dramatic theater with the motion picture camera opened entirely new problems and opportunities for filmmakers.

The differences between theater and film are basically twofold: First, the film set must be lit not merely for dramatic effect but for dramatic effect *as seen on film.* The difference can be significant. Much of the furor over colorization of old black-and-white films is based on the fact that those films were lit and staged for black-and-white showing; turning them into color films often ruins the visual impact that the director was trying to establish.

The second difference is that while a stage play may have a total of six to 12 scenes, the typical motion picture has many dozens of shots, each of which must be individually lit.

A day on a motion picture sound stage usually begins with the director, camera operator, and lighting crew "dressing" the set with light, much in the same way that a theater lighting director sets up the stage lights. The studio lighting will depend, in part, on the angles that the camera will use in photographing the scene. Only when the camera angles and lighting are firmly decided does the call go out for the actors.

"Bring in the talent," the director orders. Some of the less kindlier ones have been heard to command, "Bring in the beef."

What we see when we view a film or TV program is a flickering show of lights. Very early in the days of motion pictures, directors learned that they could do tricks with the camera that made impossible things appear to happen on the screen.

This opportunity, which we now know by the term *special effects,* originated in the mind of a magician.

Georges Méliès was a professional stage magician and manager of a Paris theater when he saw his first motion picture in 1895. He quickly realized the opportunity that film offered for tricking the eye and built a studio for himself where he produced the first special-effects films.

He invented many techniques that are still used today, including slow motion, fade-outs, double exposures, and stop motion. For example, he would photograph an actor running to the edge of a cliff. Stop the camera. Substitute a dummy for the actor and photograph it falling down the cliff and landing *splat* on the

Georges Méliès combined visual humor with special effects in his 1902 film A Trip to the Moon. *Here the intrepid French explorers' rocket hits the Man in the Moon squarely in the eye.*

ground below. Stop the camera. Put the actor exactly where the dummy landed and photograph him leaping to his feet and running away. The audience saw a man run to a cliff, fall all the way down to the ground below, jump up unhurt, and run off again. *C'est un trompe l'oeil, non?*

Interestingly, one of Méliès' best-known films was a very broad takeoff on Jules Verne's science-fiction novel *From the Earth to the Moon.* Using a variety of special effects and a chorus of cuties from the Folies Bergère, Méliès' *A Trip to the Moon* (1902) was funny—and the progenitor of the science-fiction special-effect extravaganza.

Over the years, special effects and the science-fiction film have grown up together. Innovative filmmakers have shown us giant apes, invisible men, monsters of all descriptions that regularly rampage through Japanese cities, enigmatic black monoliths that float in space near the planet Jupiter, gigantic interstellar ships engaged in furious space battles, and much, much more. Méliès spent 10,000 francs making *A Trip to the Moon.* The 1977 *Star Wars* cost more than $20 million—most of it spent on special effects.

From time to time, producers have tried to make three-dimensional motion pictures. The heyday of 3-D movies was in the 1950s, when audiences donned cheap cardboard glasses to watch various dangerous objects jabbing into their faces. Two nearly superimposed images were projected on the screen, one in a red tint and the other in blue. The 3-D glasses contained a red filter over one eye and a blue filter over the other, so that each eye saw only one of the two images on the screen. The brain put these two overlapping images together to form a stereoscopic picture.

It was not terribly satisfactory, and 3-D movies are still not very popular. Holograms—three-dimensional images made with lasers—have been tried, but they too have their drawbacks. The latest attempt to produce 3-D movies comes from television manufacturers such as JVC and Toshiba. This technique uses superimposed images on the TV screen and liquid-crystal eyeglasses that blank out each eye 60 times per second, in synchronization with the TV screen's scanning rate. Toshiba plans to market such a stereo video camera-recorder for home use.

In rock concerts the spectacular lighting effects are so important

to the audience's response that "the LDs [lighting directors] are in fact the concert 'directors'—we direct the audience where to look, we produce the visual picture," according to James Moody, who has been lighting director for many of the nation's top rock groups.

A live rock concert's lighting equipment is just as important as the sound gear. The rock group Genesis travels with 26 Vari-Lits that can rotate 360 degrees, tilt up and down 270 degrees, and change to any of 60 colors in a tenth of a second. Plus a dozen aircraft landing lights. Plus 20 additional stage lights. The lights are mounted on six triangular light bars that are hung from a nine-ton, 157-foot steel truss that comes with its own hoists and servomotors to move the lights. The whole assemblage is controlled by a small electronic computer.

Much of the emotional impact of a live rock concert comes from the sheer overwhelming power of the sound coming from the amplifiers. A Bruce Springsteen concert, for example, blasts the audience with 380,000 watts of high-power amplification.

Motion picture experts can recreate scenes from anywhere in the world—or out of it—in their studios.

221

Rock concerts rely on spectacular visual effects as well as overpowering levels of sound.

But lighting effects are almost equally important. Michael Jackson concerts have featured actual fireworks and purposefully blown-out stage lights; the performer was burned during the filming of a TV commercial when the explosive effects set his hair afire. Madonna and other groups regularly use hundreds of lights, including powerful aircraft landing lights and projection lamps, to create stunning visual effects. A rock show can have more than a thousand lighting cues that are programmed into as many as six personal computers.

What would Edison have said about all that? Would he enjoy the light show of a modern rock concert or be appalled by it? One thing is certain: The music would not matter very much to him. Edison was quite deaf.

15

Sword of Light

IT BEGAN, as many things do, with a science-fiction story:

> Suddenly there was a flash of light. . . . At the same time a faint
> hissing sound became audible. . . . Forthwith flashes of actual
> flame . . . sprang from the group of men. It was if some invisible
> jet impinged upon them and flashed into white flame. It was if
> each man were suddenly and momentarily turned to fire. . . . It
> was sweeping round swiftly and steadily, this flaming death, this
> invisible, inevitable sword of heat.

That is how H. G. Wells described the heat-ray weapon used by
Martians against Earthlings in his book *The War of the Worlds*,
published in 1898.

Although such a deadly weapon still belongs in the realm of
science fiction, today we would call it a laser.

The laser is a device that produces light: extremely intense,
brilliant light, brighter than the Sun. Laser beams are so intense
that a device of a mere few watts of output power can send out a
beam that can reach the Moon and bounce back to be detected by
equipment on Earth, a round-trip journey of nearly half a million
miles. No arc-lamp searchlight, no matter how powerful, could
do that.

It would be terribly disheartening if the only uses for lasers
were for warfare and destruction. In fact, while lasers do have
many military applications, they are also being used in medicine,
communications, research, entertainment, and a host of peaceful,

beneficial ways. Wells' sword of heat may one day protect us against ballistic missiles carrying nuclear bombs, but right now—today—laser "swords" are being used by surgeons to restore sight to the blind. In factories, laser beams are used for welding, cutting, measuring, and other tasks.

Laser disk recordings are revolutionizing the home entertainment industry. Tiny lasers, the size of a thumbnail, are sending telephone and television signals across continents and oceans at the speed of light. Lasers guide huge tunneling machines; their steady beams of light keep the machines moving in precisely straight lines.

Lasers can make three-dimensional pictures, called *holograms*. And lasers are being used in an enormous variety of research tasks, where their energetic beams of light can probe individual atoms and return information that was impossible to obtain before.

The laser is a triumph of the quantum theory of light. Newton, Maxwell, not even Einstein could have invented the laser; they simply did not know enough about the nature of light, although Einstein might have done it if he had not rejected the quantum theory.

The first successful laser was produced in 1960 at the Hughes Research Laboratories in Malibu. The word *laser* is an acronym based on the phrase Light Amplification by Stimulated Emission of Radiation. To understand what that phrase means we will have to look into the nature of light once more. Before we do, however, we should make clear the differences between lasers and other sources of light, such as stars, flames, or electric lamps.

There are four such differences: intensity, directionality, monochromaticity, and coherence.

1. *Intensity:* Laser light is more intense even than the light of the Sun. The Sun emits roughly 7,000 watts (seven kilowatts) from each square centimeter of its surface, the equivalent of the energy from seventy 100-watt bulbs coming out of an area the size of a postage stamp. Yet lasers have produced bursts of energy that contain more than a billion watts (a million kilowatts, called a *gigawatt*) in a beam of about one square centimeter in cross section. Laser beams have been focused down to tiny spots where their concentrated energy has amounted to hundreds of gigawatts per square centimeter.

Inventor and invention: Theodore Maiman holds the world's first laser, which he created at the Hughes Research Laboratories in Malibu in 1960. Ruby laser crystal is inside housing on right; helical flashlight that "pumped" light energy into the laser is on left.

(Remember, though, that the Sun is almost a million miles in diameter and emits a stupendous amount of energy from all those square centimeters. But lasers are brighter; they put out more energy per square centimeter than the Sun does.)

2. *Directionality:* Lasers can be made to emit very narrow, pencil-thin beams of light. Such beams spread very little.

Take a flashlight outdoors with you tonight and aim it at the Moon. How far does the beam go before fading out of sight? Yet lasers of only a few watts have bounced their beams off the Moon's rugged surface. They can do this partly because the laser beam is so narrow that it does not spread out and dissipate; the beam can carry much farther because it does not waste its energy by spreading. Laser beams reaching the Moon generally have spread to a circle of only a few miles' diameter.

3. *Monochromaticity:* That large word means "one color." We have seen that light from the Sun or from a flame or artificial electric lamp contains many wavelengths of electromagnetic energy. We call sunlight white light because it contains all the wavelengths of the visible spectrum. Incandescent lamps spread their energy over a broad range of wavelengths, and even fluorescent lamps emit light at several different wavelengths.

A laser emits light of one color only; its beam contains only a very few wavelengths. Some lasers have been made to emit a single wavelength of light. If you think of sunlight as the music that a full symphony orchestra makes, a laser's light is like the voice of a single violin—in some cases, a single violin playing only one note.

A laser's light is of one color and usually a very pure color as well. There are many different types of lasers, and lasers have been produced to emit light of every color of the rainbow. Some lasers emit light that is beyond human visual perception: infrared and ultraviolet. Researchers are now trying to make a reliable X-ray laser.

Some types of laser can be tuned to emit different colors. While they may be able to run the gamut from deepest red to deepest violet, they still emit only one color at a time.

4. *Coherence:* Laser light is coherent. That is, the waves of light coming out of a laser are all in step.

Picture a seaside scene, with surf curling up onto a sandy beach. Each wave is somewhat different from the others; some are big and steep, others are mere ripples. Sometimes they come spaced closely together; at other times there is a long wait between one wave and the next.

Now imagine a beach where all the waves are exactly alike. All the same height, all the same direction, always the same spacing between them. That is how the light waves emitted by a laser are.

Ordinary light sources produce light waves of many different types, jumbled together like an auditorium full of frenetic rock dancers. Lasers produce light that is precisely spaced and lined up, like a marching squad of West Point cadets. This quality of laser light is called coherence, and laser light is said to be coherent.

Now we can look more deeply into the nature of light and see how the laser was invented.

Einstein was there at the beginning. His 1905 paper on the photoelectric effect, which earned him his only Nobel Prize, convinced most physicists that the quantum theory was the best description of light. They began to picture light as consisting of tiny packets of energy called photons. Ironically, Einstein later rejected the quantum theory because he was unwilling to accept the consequences of the uncertainty principle. "God does not play dice with the universe," he would mutter.

In 1917 Einstein became interested in the way a gas will absorb and give off energy, a problem similar to the one that Planck was looking into when he first hit upon the quantum idea. Einstein found that there are three processes involved in the way a gas absorbs or gives off energy.

1. *Absorption.* This happens when a quantum of energy—a photon—is absorbed by one of the atoms in the gas. The atom gains energy. One of its electrons, having absorbed the energy of the incoming photon, jumps to a higher orbit around the atom's nucleus. As we saw earlier, it is said to be in an excited state.

2. *Emission.* When the excited atom's electron hops down to a lower-energy orbit, it gives off a photon of precisely the same wavelength as the photon it absorbed. This process is called *spontaneous emission.* Spontaneous, because the atom de-excites itself without any outside forces acting on it to do so.

In these two processes we have the workings of the fluorescent lamp. The glass envelope of the lamp contains a gas such as neon,

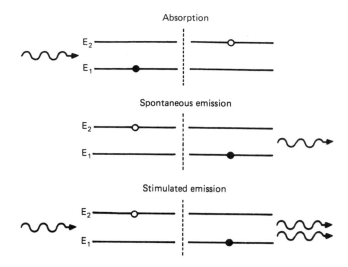

Three ways that atoms absorb and emit energy. Level E_1 represents an atom in a low-energy state; E_2 is a high-energy state. The wavy lines represent photons of light energy. Absorption and spontaneous emission are commonplace in nature. Stimulated emission is the basis of laser action.

mercury, or sodium vapor. When an electric current is applied to the lamp, the energetic electrons of the current excite the gas atoms; their orbital electrons absorb this energy, then hop down to less-energetic orbits, giving off photons in the process. These photons, usually in the invisible ultraviolet region of the spectrum, excite the chemical phosphor that coats the inside of the glass tube. It is the phosphor's glow that provides the light we see.

The third type of process Einstein found (and the one that led to the laser) is:

3. *Stimulated emission.* Einstein recognized that there must be times when an atom is already in the excited state and it is struck by another photon. In this case, he predicted, the atom will not get more excited. Instead, its excited electron will drop back to a lower-energy orbit and give off a photon. Moreover, the incoming photon will not be absorbed at all but will continue on its way out of the atom. So there will be two photons leaving the atom, where only one was coming in.

The phenomenon of stimulated emission was an interesting curiosity in 1917. It hardly ever happened in the real world. Excited atoms almost always drop back to their lowest-energy state in times of a few millionths of a second. They produce spontaneous emission and very rarely remain in the excited state long enough to give stimulated emission a chance to occur.

For nearly half a century stimulated emission was considered interesting but of no practical use. Yet those words yield two letters of the acronym "laser." Even as late as 1950 no scientist was trying to invent the laser; no one had even conceived that such a device could exist. Only a handful of science fiction writers dealt with "ray guns." And most scientists told them they were crazy.

By 1950, however, many scientists were searching for ways to make better radars. American physicist Charles Townes was interested in microwave radars. Microwaves are electromagnetic energy with wavelengths of a few centimeters' length, shorter than the radio waves used for earlier radars and only slightly longer than infrared. Certain wavelengths in the microwave region can excite molecules of water; microwave energy of the proper wavelength can make water boil. This is how microwave ovens cook food: by boiling the water inside the meat or vegetables. One early model of microwave oven was called the Radar Range.

Here we will deal with microwave radars, however, not cooking.

The heart of a microwave radar transmitter is a metal box, or cavity, called a "resonator." Like the hollow box of a violin or guitar, in which sound waves reflect back and forth and then emerge in the proper form to please the ear, a radar's resonator cavity reflects electromagnetic waves back and forth, makes the waves oscillate.

With stringed musical instruments, to make higher notes (which are shorter wavelengths) the instrument must be made smaller. A violin is smaller than a bass viol for that reason. The same principle holds true for electromagnetic waves. To makes radars of smaller wavelengths, as Townes wanted to do, the resonator cavity had to be made smaller.

Townes hit on the idea of using atoms as resonators. His first steps in that direction, undertaken at Columbia University, were to use the molecules of gaseous ammonia. Meanwhile, Joseph Weber at Catholic University and A. M. Prokhorov and N. G. Basov at the Lebedev Institute in Soviet Russia were also looking into the possibilities of using ammonia molecules as resonators.

Townes succeeded first. Using microwave energy at a wavelength of 1.25 centimeters, Townes and his assistants were able to make ammonia molecules resonate. When ammonia gas is bathed in microwave energy at that frequency (24,000 megahertz) the ammonia molecules become excited. Under normal conditions the ammonia molecules would quickly fall back to their lower-energy state, undergoing spontaneous emission in the process and emitting photons of 24,000 megahertz.

Under normal conditions, some of the ammonia molecules would be absorbing incoming microwave energy while other molecules would be emitting energy at exactly the same frequency. This is called an *equilibrium* condition. The situation would be something like a normal day at a bank, with some customers depositing money and others withdrawing.

Under equilibrium conditions, while there is a considerable amount of energy being transferred within the gas, the energy is simply going from one molecule to another. It is too disorganized to be of much use.

Townes, however, wanted to arrange conditions so that a large number of the ammonia molecules became excited at the same time and emitted photons together. If this could be done, the energy output would be highly organized and useful. The situation would be something like a run on the bank, where everybody wants to withdraw money at the same time. Fortunately, conditions can be arranged so that fresh supplies of energy are fed into the ammonia gas, so that the molecular "bank" will not run out of energy "money."

The Columbia team was able to produce a nonequilibrium state in the gas, where the molecules remained excited long enough to arrange the proper conditions for stimulated emission, rather than spontaneous. A cascade of photons at the microwave wavelength of 24,000 megahertz came out of the gas, and the world had seen its first maser—Microwave Amplification by Stimulated Emission of Radiation.

Masers have become important in radars, communications, and any applications where one needs extremely sensitive radio receivers that can be tuned to very precise wavelengths. Astronomers who use huge radio telescopes to search for possible signals from extraterrestrial civilizations use masers to help tune in on the very faint signals coming from distant stars.

If microwaves can be generated by stimulated emission, why not waves of visible light? Townes began working with his brother-in-law, Arthur L. Schawlow, who was then a research physicist at Bell Labs. They began dreaming of an "optical maser."

They needed to find a type of molecule or atom that could be pumped up to an excited state and then held there without undergoing spontaneous emission. At that point a few photons of the proper wavelength would make the entire population of excited atoms undergo stimulated emission, and a cascade of photons would be emitted.

The trick of getting atoms or molecules into the excited state and then holding them there is called *population inversion.* Under normal conditions the atoms tend to remain at the lowest possible energy level, which physicists call the *ground state.* Excited atoms quickly fall back to the ground state—normally.

In 1958 Townes and Schawlow published a paper outlining their theoretical understanding of how to make an optical maser. Two years later the first laser began to shine.

It was the work of Theodore Maiman, who built the first laser at the Hughes Research Laboratories, perched on a lovely hillside that overlooks Malibu's famous beach.

Maiman chose an artificial ruby as the working medium for his laser. The ruby crystal is composed of a mixture of aluminum oxide and chromium oxide. The more chromium atoms present, the deeper the red color of the ruby. Maiman chose a pale pink ruby that contained about half of 1 percent chromium. He wanted to excite the chromium atoms and then get them to give off photons by stimulated emission.

Maiman hit upon the idea of using light itself to pump up the chromium atoms to a nonequilibrium state of excitation. He used a powerful flash lamp for this purpose. His ruby was a tiny cylinder, four centimeters long and half a centimeter in diameter. The ends of this little rod were polished flat and partially silvered so that they became mirrors that would reflect light up to a certain intensity. If the light became strong enough, these partially silvered mirrors would allow the light to pass through them and into the outside world.

A helical flash lamp was wound around the ruby rod, coiling about it so that its light would strike the rod's entire surface.

When the lamp flashed it broadcast a powerful burst of light in all directions. Some of this light energy went into the ruby and excited the chromium atoms—the first step toward laser action.

The chromium atom can reach two excited energy levels under these conditions. Once it has reached either of them, it immediately falls back to an intermediate energy level—not the ground state, but an excited state that is about halfway between the ground level and the highest energy level it can attain.

The chromium atoms tend to stay at this intermediate energy level for a few thousandths of a second. This is a *long* time in the frenetic world of atoms. At this point they are said to be in a metastable state, a condition that is only partially stable. While in this metastable state, a few of the chromium atoms will spontaneously emit photons and drop back down to the ground state. These spontaneously emitted photons are the first few pebbles that start the landslide.

The photons strike other chromium atoms that are still in their excited metastable state. Since the photons are exactly at the precise wavelength necessary, they produce stimulated emission. Each time a photon strikes a metastable chromium atom, another photon of the same wavelength is produced. Now the two photons go on to strike two more metastable chromium atoms.

The photons are moving in all directions, at random—up, down, right, left—but always in straight lines. Many of them escape from the ruby rod right away, but a number of them happen to be moving in a direction along the axis of the rod. They get as far as the mirrored end of the rod and are reflected back in the direction from which they came.

So they go streaming down the length of the rod again, striking more metastable atoms along the way and causing more stimulated emission, making a bigger and bigger cascade of photons surging back and forth between the mirrored ends of the rod.

All this takes place in a few millionths of a second, of course. When the cascade of photons reaches a high enough intensity it passes through the partially silvered mirror at one end of the ruby rod and comes out as a pulse of brilliant laser light.

Now we can understand why laser light is so different from all other light sources.

Intensity. Laser light is brighter than any other type of light because things are arranged in a laser so that a very high percentage of the excited atoms contribute photons to the light output. In ordinary light sources—for example, the flash lamp that pumped the first ruby laser—only a few atoms are giving up photons at any particular instant of time. In the laser, the atoms have been organized much more efficiently.

This is akin to what happens when a crowd of people yell at a

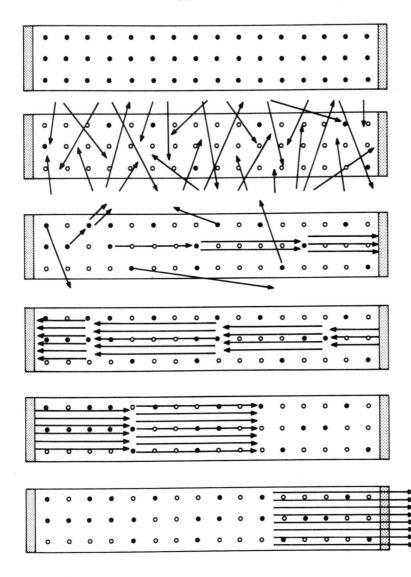

How a laser works. The white circles are excited atoms; the arrows indicate the paths taken by photons. As the photons are reflected back and forth by the mirrored ends of the laser cavity, excited atoms give off more and more photons through stimulated emission, until a cascade of photons is emitted from the laser in unified, coherent, single-direction waves.

football game. If the people holler at random, individually, there is a steady hum of noise. But if the cheerleaders get them all to shout in unison, the sound can rock the stadium.

Directionality. Here again, conditions are arranged in the laser to make the output beam very narrow. This is a consequence of the shape, or geometry, of the laser. The photons that come out as laser light are those that have traveled back and forth along the axis of the laser cavity. (In the case of the ruby laser, the cavity is the ruby rod. In other types of lasers, as we shall see, other kinds of cavities exist.)

The photons that move off-axis are lost, but those that *do* move along the axis build up a powerful cascade and come out of the laser cavity in a narrow beam that spreads very little and thus can carry a very long way. In this regard a laser is like a long-range precision rifle, while other light sources are like shotguns.

Monochromaticity. Only photons of the same wavelength can be emitted from the laser. Well, almost. Actually, most lasers emit light in a narrow band of wavelengths. Maiman's first ruby laser emitted light in a very tight cluster of wavelengths grouped around the 694.3 nanometer line. The laser is much more monochromatic than any other light source.

Coherence. If some fantastic slow-motion device could allow us to see individual waves of light as they come out of a laser, we would see that the waves emerge all in step; physicists say they are *in phase.* They are all of the same size, the same shape, and they all have the same spacing between them; they are all perpendicular to the direction in which they are moving. This is called *spatial coherence.* They are also equally spaced, so that the time between one wave crest and the next is always the same; this is called *temporal coherence.*

It is a truism in science—in fact, it is true of all human endeavor—that once something new has been accomplished, no matter how difficult it may have been to achieve, others rush in and do the same thing better, more easily, more efficiently. As the entertainer Jimmy Durante used to say, "Everybody wants to get into the act!"

Maiman's breakthrough with the pulsed ruby laser opened the floodgates. Throughout the 1960s and 1970s, right up to the present day, scientists and engineers all around the world have produced hundreds of different types of lasers, using all sorts of material as the "lasing" medium. Laser action has been produced from solids, such as the original ruby crystal, from gases, liquids, semiconductors very much like the chips used in computers, and even ordinary air.

A high school student from Boston's Roxbury ghetto built a homemade laser out of the glass tube from a burned-out fluorescent lamp, the compressor motor from a junked refrigerator, and a bit of ingenuity. His was a gas laser, and the gas that "lased" was a mixture of carbon dioxide, nitrogen, and water vapor, with some traces of other gases mixed in. Where did he get this gaseous mixture? From his own breath. He exhaled into the tube.

In general, lasers can be classified either by the material that is used to produce laser action or by the method used to excite the active atoms or molecules. The material can be solid, liquid, or gas. The method of exciting the material, usually called the pumping method, can be optical (using light, as Maiman did), electrical, or even chemical.

There is one other way to classify lasers: by the nature of their output beams. Some lasers are inherently pulsed devices. The beams they emit come in short bursts of light. I worked at the Avco Everett Research Laboratory in the 1960s, where our scientists produced a laser that used nitrogen gas as its working medium. It emitted ultraviolet light, invisible to the unaided eye, in pulses that were only a few nanoseconds long. A nanosecond is one-billionth (10^{-9}) of a second. There are as many nanoseconds in one second as there are seconds in 32 years. Many solid-state lasers are pulsed devices, while most gas lasers emit a continuous beam of light and are called *continuous-wave* (CW) lasers. Plate 21 (following page 238) shows a modern laser at work.

Solid-state lasers employ crystalline materials as their laser medium. The first laser used artificial ruby. Lasers have been made of glass "doped" with atoms of metal that produce the laser action. YAG lasers use synthetic garnets that contain the rare earth elements yttrium and aluminum, together with a trace of neodymium, another rare earth element. YAG stands for yttrium-aluminum garnet, although the laser action comes from the neodymium while the other elements serve merely as a host for the active neodymium atoms.

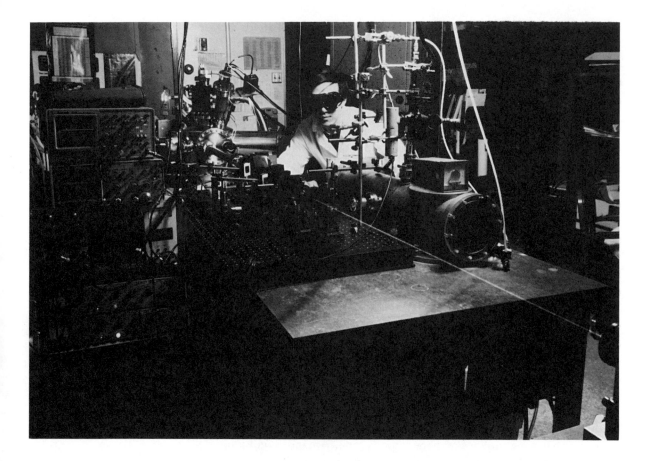

An excimer laser, used at the Naval Research Laboratory to study the reactions of gases with various kinds of solid surfaces. Excimer lasers are based on principles that most chemists would have considered impossible a generation ago.

Many other kinds of solid-state lasers have been built. They are almost always optically pumped; that is, the input energy comes from a flash lamp, similar to the technique used by Maiman.

Remember the organic dyes that we looked at in Chapter 12? It should come as no surprise that these natural pigments are capable of laser action. While they are usually solid at room temperature, for laser action they are dissolved in an organic compound such as alcohol to form liquid lasers. Such dye lasers can be tuned to produce a broad variety of output wavelengths. Tunable liquid lasers can change the color of their beams with the twist of a dial.

Gas lasers come in many varieties, from the low power helium-neon types that are often used for classroom demonstrations to mammoth carbon dioxide lasers that emit megawatts of energy in beams powerful enough to shoot down an airplane or missile. While some gas lasers are pulsed devices, most emit a continuous beam. Most gas lasers are pumped electrically; electrical energy excites the gas atoms directly. As in every type of laser, the heart of the gas laser is the so-called laser cavity.

For most gas lasers the cavity is a glass tube filled with the "lasing" gas. Some types of gas laser, however, look more like a wind tunnel than a fluorescent lamp. The gas is blown through the cavity, often at supersonic speed, then it is either recycled to be used again or allowed to blow away altogether. Whatever the nature of the cavity, however, it always has mirrors on each end to reflect the cascade of photons building up inside it and some means of allowing the laser beam to get into the outside world—a window, in other words. In the first ruby laser, as we saw, the mirror and the window were merely a partially silvered end of the ruby cavity.

Semiconductor lasers are also electrically pumped. You have seen light-emitting semiconductors in digital clocks, pocket calculators, and other electronic appliances. The light comes from a semiconductor device called a *light-emitting diode,* or LED. Such diodes can also be made to produce laser light. Diode lasers are so small that you could hold dozens of them in your hand. They are also very efficient and reliable, capable of running continuously for hundreds of thousands of hours or more.

As you might suspect, tiny diode lasers are not very powerful. Nor need they be for most of their applications. Top output powers for semiconductor lasers are a few watts CW and pulses of a few kilowatts. As we shall see, such lasers are used principally in communications, where power output is not so important as reliability and efficiency.

Then there are the chemical lasers, where laser action is caused by mixing two or more energetic gaseous elements such as hydrogen and fluorine. While other kinds of lasers need an outside source of energy to excite the working medium, in chemical lasers the working medium itself contains enough chemical energy to trigger laser action. Mix the chemicals, usually by burning them together, and the gas resulting from this combustion can contain an excited population of atoms that can undergo laser action. The down side of this situation is that once laser action has been

PLATE 19
van Gogh's room at Arles, a study in color.

PLATE 20
Although the development of photography took away straight pictorial realism from the creative painter, artists soon found that they could convey reality in ways that photography could not. Claude Monet's La Rue Montorgueil Adorned with Flags *shows a scene we can all recognize, but in a way that we could not see for ourselves.*

PLATE 21
A tiny man-made star is created for a brief fraction of a second when the 24 beams of the intense Omega laser strike a micropellet of deuterium-tritium fuel and trigger a burst of nuclear fusion.

PLATE 22
Gemstones come in many sizes, shapes, and colors, but they all share one quality in common: the ability to dazzle the eye with a brilliant display of refracted and reflected light.

(Captions continue on page 239.)

238

PLATE 19

PLATE 20

PLATE 21

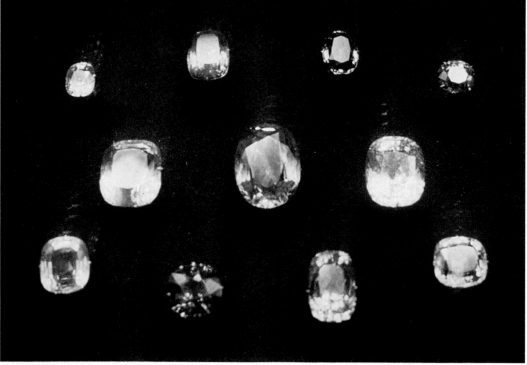

PLATE 22

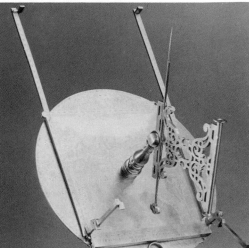

PLATE 23

PLATE 24

PLATE 25

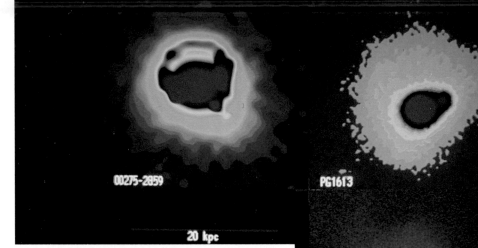

20 kpc

PLATE 27

PLATE 26

PLATE 23
An astrolabe-type of instrument, used to measure the angular positions of stars. Built in 1557 in Florence, this kind of instrument was the apex of astronomical apparatus prior to the invention of the telescope.

PLATE 24
The Sun in all its fiery splendor, as seen from the Skylab space station. Each dot on the Sun's surface is larger than the planet Earth.

PLATE 25
Stars are formed out of clumps of gas and dust, as these protostars are in the process of doing. When the temperatures at their cores reach the level necessary for hydrogen fusion, the protostars will begin to shine and become true stars.

PLATE 26
Supernova 1987A.

PLATE 27
False-color images of quasars, enigmatic objects that may be at the very edges of the observable universe, more than 10 billion light-years from Earth.

produced, that batch of chemicals cannot easily be made to lase again; the chemical laser must be constantly supplied with fresh chemicals.

While the first chemical lasers all emitted infrared beams, in 1987 a research team at the Georgia Institute of Technology produced a chemical laser that emits a bright green light.

Chemical lasers are among the most efficient, typically converting 20 percent or more of their chemical energy into laser light. Most gas lasers convert less than 10 percent of their input electrical energy into laser output.

One of the more recent developments is the excimer laser, which is something of a cross between an electrically pumped gas laser and a chemical laser.

While lasers in general owe their existence to the quantum revolution that shook physics in the first half of the twentieth century, the excimer laser owes its existence to a small revolution that rattled the world of chemistry two years after the first laser was built.

Until 1962 it was an axiom of chemistry that the family of elements called the noble gases were chemically inert. That is, they would not form chemical compounds with other elements. The noble gases, in order of their atomic weight, are helium, neon, argon, krypton, xenon, and the radioactive radon. You breathe a smattering of them every time you inhale; they form trace elements in the Earth's atmosphere.

There were perfectly valid reasons for supposing that the noble gases do not react with other elements. But scientists are curious and inventive. In 1962 English chemist Neil Bartlett showed that xenon could link chemically with fluorine—which is one of the most active elements of them all. If xenon is "noble" in the sense that it tries to remain aloof from all other elements, fluorine is a used-car salesman who is running for mayor.

In an excimer laser, an electron gun fires a beam of very energetic electrons into a mixture of a noble gas and a halogen (chlorine, fluorine, bromine, or iodine). Suppose the mixture is xenon and fluorine. The energetic electrons cause the two gaseous elements to combine to form an excimer, which is a compound that can exist only in the excited state. So electrical input energy induces a chemical combination that yields an excited molecule, called an excimer. When the excimer is stimulated to emit a photon, laser action begins.

There are several other types of lasers, including the X-ray laser, which is pumped by the energy of a nuclear explosion! The X-ray laser is the pet brainchild of the controversial Edward Teller, father of the American hydrogen bomb. Teller foresees X-ray lasers, which he called "third-generation nuclear devices,"* being used in space as part of a "Star Wars" type of strategic defense system. The nuclear explosion would energize the X-ray laser before it is destroyed by the blast, and the laser would beam X-rays at oncoming ballistic missiles of sufficient intensity to destroy them.

Many people who support the Strategic Defense Initiative feel that the nuclear-pumped X-ray laser is unnecessary and unwanted. However, if physicists learn enough to produce X-ray lasers without the need for a nuclear explosion to energize them, such lasers would find enormous peacetime applications in industry and medicine.

While X-ray laser experiments have been part of the underground nuclear bomb tests conducted in Nevada, researchers at Lawrence Livermore National Laboratory in California carry out X-ray laser experiments using an extremely energetic laser system called Nova. The Nova system does not produce X-rays directly. Its laser beam strikes a foil of selenium metal with a nanosecond pulse of some 600 trillion watts. This excites the selenium atoms to radiate X-rays.

The Lawrence Livermore team is studying the feasibility of producing X-ray holograms. As we will see in the next chapter, holograms are three-dimensional pictures made with lasers. Three-dimensional X-ray pictures would be an important medical breakthrough, of course.

Lasers began as science fiction, and for many decades any thought of a device that produced beams of energy was dismissed as Buck Rogers fantasy. Even when the first lasers were demonstrated, most people scoffed at the idea of "death rays."

Today lasers of various types produce beams of light all across

---

*First-generation nuclear devices are fission bombs of the type that destroyed Hiroshima and Nagasaki; second-generation are the fusion-based hydrogen bombs.

the visible spectrum and far into the infrared and ultraviolet as well. Today lasers are being used in thousands of ways in factories, laboratories, and hospitals. They may become weapons of defense, to protect against ballistic missiles carrying nuclear bombs. They may also become death rays that pinpoint cancerous growths and destroy them.

The uses of the laser seem to be virtually limitless.

16

Bright Beams of Laser Light

TOOLS ARE basically extensions of our bodies.

A telescope is a device for extending our vision over astronomical distances. A lever is a device for increasing the force our muscles can apply. A shovel allows us to dig better than we could with our bare hands.

The laser is a tool that can deliver energy over a distance. In some cases the energy may be tiny, fractions of a watt, and the distance microscopic. In other cases the energy may be many megawatts and the distance thousands of miles.

Lasers are being used in factories to cut metal and in surgical theaters to remove cataracts. Laser beams have measured the distance to the Moon to within a few inches and have been proposed as weapons that will shoot down ballistic missiles.

Lasers have been used by physicians at the San Francisco Heart Institute to "weld" human blood vessels together, replacing sutures and staples. Fifteen such procedures were done there in 1987. Laser beams may soon also be used in angioplasty—melting away the constricting plaque inside blood vessels that can lead to a heart attack or stroke.

The light from a laser can also protect blood stored in blood banks from deadly viruses, including the HIV virus that causes AIDS. A team of Texas researchers (from the Baylor Research

Lasers are used to cut and machine metal. Fully automated laser machining systems remove metal faster than mechanical systems and allow much more precise control of the process.

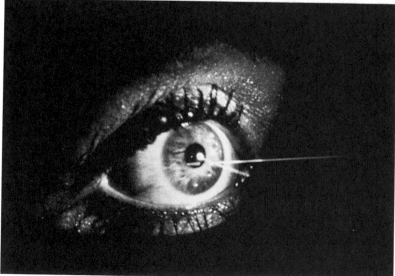

Beams of laser light are used to diagnose eye diseases, to seal ruptured blood vessels inside the eye, even to remove cataracts and reshape the cornea.

Foundation, Southern Methodist University, and the Southwest Foundation for Biomedical Research) have developed a method for treating stored blood with a light-absorbing dye, and then exposing the blood to a low-intensity laser beam. The researchers claim a 100 percent kill rate for viruses that cause measles, herpes simplex type 1, AIDS, and others.

Artisans use laser beams to etch metal and burn patterns into wood, creating the new art form of laser engraving. Credit cards will one day carry microscopic data dots that will be read by laser beams in place of the magnetic strips they now have. Such "smart cards" will be able to check each charge against your bank balance and prevent overdrafts; they will also be much more difficult to forge or alter.

Lasers check the prices of your groceries at supermarket checkout counters and play the music of your compact disk records. In laboratories around the world beams of laser light probe and measure everything from earth tremors to the vibrations of individual gas atoms. Scientists can now observe chemical reactions *as they take place,* even though the intricate interplay among the atoms happens in picoseconds (trillionths of a second). Lasers that emit pulses only a hundred-trillionth (10^{-15}) of a second long yield "snapshots" of the submicroscopic dance of the atoms as they bond themselves together or break apart. Construction crews use lasers for measurements; so do factories; so do the gunners of the world's armies and navies.

Most of these uses for lasers stem from the fact that beams of laser light carry not only energy. They can carry information.

Modern communications systems transmit information electronically. When I telephone Arthur C. Clarke at his home in Sri Lanka, for example, my voice is transformed into an electronic signal that is transmitted to a communications satellite orbiting 22,300 miles above the equator. The signal is then beamed down to Arthur's telephone. (This is a very appropriate thing to do, since Clarke invented the idea of communications satellites in 1945.)

The signal that carries my voice, and Arthur's reply, is a set of radio waves, electromagnetic energy in the radio-frequency part of the electromagnetic spectrum. Typically, telephone signals are transmitted in the region of 1,000 to 4,000 hertz. (You recall that one hertz is one cycle per second.)

At the heart of the automated checkout counter at many supermarkets is a diode laser that "reads" the product information labels on the goods being checked out.

Television signals must carry much more information; an electronic picture is equal to much more than 10,000 electronic words. Television signals use frequencies of 50 to 500 megahertz. A 50-megahertz TV signal, therefore, requires as much signal-carrying capacity as 12,500 to 50,000 telephone conversations. Communications engineers refer to *bandwidth:* the range of frequencies needed to transmit or receive a signal. Television requires a much wider bandwidth than voice-only radio or telephone signals.

Think of the waves of an electronic signal as a set of spikes on which you can place a piece of information, like sticking a note on the old-fashioned spindles that used to adorn newspaper editors' desks. The higher the frequency of the electromagnetic energy, the more tightly bunched its waves are, the more information can be carried. Thus communications engineers are constantly seeking higher-frequency channels, which offer broader bandwidths. The first TV sets used VHF (very high frequency) signals. Not enough to satisfy the growing demand for television channels, so the TV industry moved into the UHF (ultrahigh frequency) region.

There is a limit to how high a frequency you can reach with radio waves. There is a limit to how much information radio waves can carry. Light waves, on the other hand, have frequencies on the order of 10 to 800 gigahertz (one gigahertz is a trillion, 10^{12} hertz). They vibrate at frequencies that are a hundred thousand to a million times greater than radio waves. Light waves can carry millions of times more information than radio waves. A single laser beam could theoretically carry 16 million TV broadcasts at once.

Lasers are being used in communications. Laser light, beamed through fiber-optic networks, is replacing much of the copper wire in telephone systems, as we will see in more detail in the next chapter.

But there is one type of communication task that only lasers can do, a new capability that could only be dreamed about before the advent of the laser: three-dimensional pictures.

It is called *holography,* a word that has an interesting history. Originally, *holograph* meant "a document written entirely by the hand of the person named as its author." The word was built from Greek roots: *hol,* meaning "entirely"; and *graphos,* meaning "to write." Thus a holographic will is one that has been written entirely by the person whose will it is.

It was the Hungarian physicist Dennis Gabor who annexed the word *holography* for application in the area of three-dimensional pictures. In 1947 he was at the Imperial College of Science and Technology in London, working with electron microscopes, a new invention in those days. He thought that it might be possible to make three-dimensional pictures out of the electron beams used by the microscope or even with X-rays. His work never bore fruit in the field of electron microscopy, but it flowered in the area of visible light once lasers came along. (And X-ray lasers may make his dream come true.)

To make an ordinary photograph you need a subject, a light source, photographic film, and an optical system. The film and the optical system are usually packaged together in the device we call a camera. Light from the source is reflected off the subject,

To make a hologram, the laser beam is split into two parts. The object beam is reflected off the subject to be photographed; the reference beam reaches the film without touching the subject.

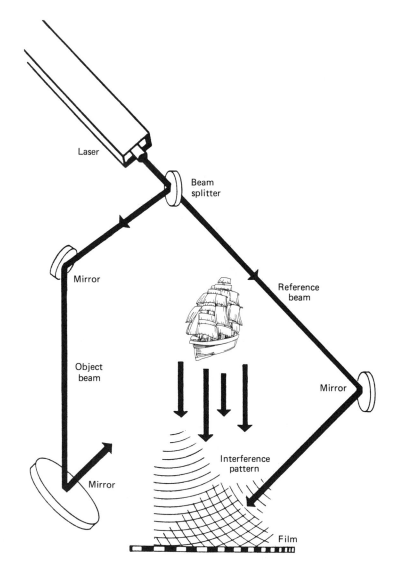

focused by the lens of the optical system, and recorded on the film in a photochemical reaction.

In holography, instead of recording the *image* of the subject on film, we record the light waves that the subject reflects. A hologram "freezes" the light waves and stores them so that we can "play back" their pattern and re-create the original subject—in three-dimensional depth, rather than the flat picture that ordinary photography yields.

While Gabor knew that holography was theoretically possible, and even experimented with ordinary light sources prior to 1960, it was evident that a source of coherent light was necessary to make a hologram. The idea was there, but it had to wait until the proper tool was invented: the laser.

The first holograms were produced in the early 1960s by Emmet N. Leith and Juris Upatnieks of the University of Michigan. They illuminated their film with two patterns of light. The laser beam was split into two parts. One part of the laser light was reflected off the subject and onto the hologram film. The other part of the beam was bounced by a mirror directly onto the film without touching the subject at all. The two beams set up interference patterns, which were recorded on the film. The film, incidentally, can be ordinary photographic film or a photographic plate in which the optically active chemicals are sandwiched between two sheets of glass.

Instead of a piece of film with an image on it, the hologram film shows only a confused swirl of interference patterns. Embedded within them is the frozen record of light waves that were reflected off the subject mixed with light waves directly from the laser. But that crazy-looking hologram contains all the information needed to reconstruct the subject in three dimensions, almost as solid and real-looking as if it were standing in front of us.

To reconstruct the image, shine a laser beam on it of the same wavelength as the original laser used to make the hologram. The subject appears to take form in midair as the laser beam shines through the film. The subject seems solid, three-dimensional. If you move around the image, you will see it shift and reveal parts that were originally hidden from view.

Actually, holographic images do not yet look quite as real as the original subjects, partly because the holographic images twinkle

Although it is difficult to convey the three-dimensional effect of a hologram in a two-dimensional reproduction, this hologram of a sailing ship is fully three-dimensional.

slightly. This is due to some of the microscopic interactions between the laser light and the air it travels through. Researchers are working to improve the quality of holographic images, and they have learned to reconstruct them using ordinary white light rather than a laser beam. But the holographic image that is so perfect and three-dimensional that it cannot be distinguished from a real person or object—that is still in the realm of science fiction, for a while.

One of the most astounding things about holography is that

every part of a hologram contains all the information of the entire hologram. Tear a piece off a hologram and you can still get a complete three-dimensional picture out of it. And out of the original hologram that is missing a piece. The image formed from the scrap is weaker, fainter than the original. But all the information is there. This means that holograms can be made very small and still carry a tremendous amount of information.

Computers that store information in holograms and use thumbnail-sized diode lasers could be made much smaller and faster than today's electronic computers. The hard disks of today's electronic computers store information in the form of tiny magnetic fields impressed on the disk. They can pack the equivalent of 600 typewritten pages into the area of a postage stamp.

A single hologram of one square centimeter, a little more than half the size of a postage stamp, could contain as many data bits as 14 cubic feet of ordinary computer memory storage. Go to the kitchen and hold a postage stamp alongside your refrigerator and you will see the tremendous advances that optical computers can offer.

Using laser beams in place of electrical current to process information in a computer will produce much faster machines. The light beams move faster and do not generate as much heat as electrical devices. *Photonic* computers can be smaller, faster, and more reliable than their electronic predecessors. Early in 1988 scientists at Bell Communications Center in New Jersey revealed a tiny fused-quartz optical device that can switch a beam of light from one optical fiber to another in less than a picosecond (10^{-12} sec), thousands of times faster than any previous electronic or even electro-optical switch.

But small size and swiftness are not the only advantages that optical computers may have. Holographic memories will bring the world's most powerful computers down to palm size. Because of this enormous information-handling capability, photonic computers might be built large enough to rival the information capacity of the human brain.

The first truly intelligent computer may work on beams of laser light. Long before that day dawns, however, beams of light from diode lasers will begin to replace the electric currents inside computers.

Bell Laboratories scientist Charles V. Shank and the laser that has produced the shortest pulses of light— 30 picoseconds (30^{-12} sec). Ultrashort light pulses are used to study subtle changes in matter, such as how electrons move in semiconductor materials.

While H. G. Wells and others have seen lasers as swords of light, they can also be used as scalpels. In fact, laser scalpels can cut more finely than the sharpest steel, and the laser beam cauterizes as it cuts, the heat of the laser beam sealing off blood vessels by searing them as it goes by. This reduces the amount of bleeding involved in surgery.

The most dramatic use of lasers in surgery comes in treating degenerative diseases of the eye itself. No other tool except light can enter the eye without injury. But since the lens and the fluids inside the eyeball are transparent to light, low-power lasers can be used as miniature scalpels to correct problems that would, untreated, cause blindness.

Diabetics are subject to a degeneration of the retina called central retinopathy, in which blood vessels proliferate on the surface of the retina. They often rupture and spill blood into the clear fluid of the eyeball, obscuring vision, sometimes totally. Using the blue-green light from a small argon laser, eye surgeons have been able to destroy the blood vessels that block the retina without damaging the retina itself. The laser beam, precisely aimed and focused in part by the eye itself, is apparently absorbed by the blood, which is flash-heated and vaporizes. The treatment is far from 100 percent effective, but there is no other method to attack the proliferating blood vessels that cause diabetic retinopathy.

Lasers have also been used to literally weld a detached retina back onto the rear wall of the eyeball and to remove cataracts that have clouded the cornea.

Because laser light can be focused down to extremely tiny spots and can be made very intense, laser beams have been used to zap lesions and cancerous growths both outside and inside the body, with a minimum of injury to the healthy tissue surrounding the diseased cells. Laser light can be carried inside the body on fiber-optic "light guides" and used to attack bleeding ulcers and even tumors in the bladder and brain.

Laser surgery has shown great promise in gynecological disorders, including problems such as heavy uterine bleeding that previously required hysterectomy. By using a laser instead of a scalpel, the trauma of gynecological surgery is greatly reduced; complications and pain are lessened. Operations that required a week's stay in hospital are now being done in a doctor's office, and the patient can return to normal activity in 48 hours or less.

Ophthalmologist Herbert E. Kaufman of the Louisiana State University at New Orleans foresees the day when laser beams may be used to alter the shape of the cornea and eliminate the need for eyeglasses altogether. The cornea, you recall, does most of the eye's focusing. Vision problems stemming from the cornea's inability to focus could be corrected by using a laser beam to delicately reshape the cornea, sculpting it a millionth of an inch in each flash to bring about the correct shape for precise vision.

Lasers are also now being used in eye examinations. The laser ophthalmoscope produces high-resolution images of the interior of the eye quickly, painlessly, and without using drugs to dilate the pupil.

Lasers are also being used in cosmetic surgery to erase birthmarks or tattoos that the owner no longer wants adorning his or her body. Low-power laser light has also been effective in speeding the healing of skin ulcers.

In the final analysis, though, Wells was right. Lasers can be used as swords, as military weapons. And they undoubtedly will be.

I have a personal bias here.

In 1966 I helped to set up a top-secret meeting in the Pentagon to tell the Department of Defense that lasers were no longer laboratory curiosities. Scientists at the laboratory where I was employed as manager of marketing had made an important breakthrough. Lasers of megawatt power were now possible.

It was an interesting meeting, that cold February morning in 1966. The meeting had originally been scheduled for a month earlier, but a blizzard buried the entire East Coast in several feet of snow, and the meeting was postponed.

In February, the nation's top experts in optics, electronics, lasers, and gas physics were gathered by DOD to listen to the presentation made by the scientists of Avco Everett Research Laboratory. The slide projector broke down, and not one of these PhDs could fix it! We had to bring in a lowly tech sergeant (who was not allowed even to glance at the slides, they were so hush-hush).

Once the recalcitrant projector was fixed, the Avco scientists showed how they had invented the *gasdynamic* laser, a laser that had already produced 10,000 watts of power in its beam and could be scaled up to produce millions of watts. At the end of the presentation, which went on for nearly two hours, I turned off the slide projector and turned on the overhead lights. There was a long moment of absolute silence in the conference room. Then the director of the Avco laboratory, Dr. Arthur Kantrowitz, muttered, "If you don't believe us, next year we'll come back and melt a hole in the wall of the building!"

They did believe us. And immediately the Pentagon began studying ways in which very powerful lasers might be used. Since I was involved in marketing the laboratory's products, I quickly saw that the gasdynamic laser's intense beams could drill through rock, could slice through the toughest steels, could weld metals. We began working with industrial groups in all these areas.

But the most fascinating possibility was to use powerful lasers, based in orbiting satellites, to destroy ballistic missiles bearing hydrogen bomb warheads. In March 1983, President Reagan announced the Strategic Defense Initiative—which the media immediately dubbed "Star Wars."

The concept of using lasers (or other exotic weaponry) to defend against ballistic missile attack has stirred enormous controversy. But despite the oratory, and no matter what your opinion of the idea may be, the physical facts are as follows.

Laser beams, at the speed of light, can strike across thousands of miles of space to destroy a ballistic missile. Picture the explosion of the space shuttle *Challenger:* That is what would happen to a missile if such a laser beam struck it while its first-stage rocket engines were boosting.

By placing such lasers in satellites, it is conceivable to patrol every square inch of planet Earth, so that missiles launched by anyone from anyplace (including the oceans) can be detected and destroyed.

Since 1945 the world has lived under the shadow of nuclear devastation. The mushroom clouds over Hiroshima and Nagasaki have darkened all our lives. Today, there are some 50,000 nuclear bombs in warheads around the world; they have an aggregate explosive power equal to some 20 billion tons of TNT—roughly four tons of TNT for each man, woman, and child on Earth.

The bright beam of laser light, the sword of heat that H. G. Wells first wrote about nearly a century ago, may make it possible to dispel the cloud of doom that hangs over us. This shining new technology may light the way for the politicians and diplomats to a world of peace.

There are enormous *ifs* involved in strategic defense, doubts and questions that deal quite literally with the life and death of each human being on Earth.

Only one thing seems certain. In the endless round of military technology, the advent of the laser has helped to tip the scales toward the side of the defense. Lasers may one day protect every one of us against nuclear holocaust—if we are wise enough to develop this technology, and the politics to control it, for peaceful ends.

Lasers can also be used to help generate nuclear energy. The

One of many suggested schemes for using lasers in a strategic defense system. The beam from a powerful ground-based laser is reflected off mirrors in space to destroy approaching ballistic missiles.

concentrated light of laser beams has already triggered a miniature man-made star that flickered for a tiny fraction of a second in a laboratory in Michigan.

Scientists know of two ways to tap the enormous energy hidden within the nucleus of the atom. The first way is nuclear *fission*, in which heavy nuclei such as uranium and plutonium are split apart to release energy. The first atomic bombs were fission bombs. A few pounds of uranium and plutonium released the explosive equivalent of roughly 20,000 tons of TNT and leveled Hiroshima and Nagasaki in 1945.

The second way is nuclear *fusion*, in which nuclei of the lightest element, hydrogen, are forced together to form helium nuclei, releasing energy in the process. Fusion is the energy source of the

256

stars. The Sun and all the stars are fusion generators, forcing nuclei to fuse deep in their ultrahot cores and releasing this energy as light and other forms of electromagnetic energy.

The hydrogen bomb is a thermonuclear fusion device where fusion happens in a violent, uncontrolled way to release much more energy than atomic bombs do. The Hiroshima and Nagasaki bombs were rated at about 20 kilotons TNT equivalent. Hydrogen bombs can release megatons of explosive power. The largest one ever exploded was a Soviet bomb of more than 60 megatons—3,000 Hiroshimas in one bomb.

Since the end of World War II physicists the world over have been attempting to harness fusion for the peaceful production of energy.

The nuclear power plants in use today are fission plants. They use heavy elements such as uranium for their fuel and produce waste products that are highly radioactive.

A fusion power plant could use water as its fuel, and its waste product would be inert helium—the stuff that fills up balloons and blimps.

It works this way.

The nuclei of four hydrogen atoms are fused together to form one helium nucleus. In this process, 0.7 percent of the matter of the original four nuclei is converted into energy. Remember Einstein's $E = mc^2$? In this case, the final helium nucleus is 0.7 percent lighter than the original four hydrogen nuclei. That amount of matter has become energy. Not a large amount, certainly. Just enough to light the stars, to keep the Sun shining for billions of years, to provide the energy that makes life possible.

Every second, the Sun converts some 4 million *tons* of its matter into the energy we call sunlight. Yet the Sun is so massive that it can continue shedding 4 million tons of material per second for at least 10 billion years. And at the end of that time it will have lost only 0.7 percent of its original matter.

Making a thermonuclear power generator, then, is a fairly straightforward matter. Take a ball of hydrogen gas about a million miles in diameter and leave it alone. Natural processes will produce a star—a thermonuclear power reactor par excellence.

That works fine for the cosmos. But physicists who want to produce a thermonuclear reactor here on Earth cannot use the

Twenty-four beams of laser light are generated in the Omega system. Peak power is 10 trillion (10^{13}) watts, approximately twenty times the electric power generating capacity of the entire United States! But the laser pulse lasts only one billionth of a second.

same brute-force methods that nature uses. They must deal with spoonfuls of material, which means they must learn how to create fusion the hard way.

The hydrogen atom is the simplest of all the elements. Its nucleus is a single positively charged proton, orbited by a single negatively charged electron. In the core of the Sun, the enormous pressure of 10 billion billion billion tons of matter pressing down produces temperatures of 20 million degrees. The electrons are stripped away from their nuclei, and the bare nuclei—positively charged protons—are forced to fuse together.

Just as on Earth, like charges tend to repel one another. The protons only come together because the enormous forces of heat and pressure overcome their electrical repulsion. While the actual processes of fusion are somewhat more complex, the net result of the process is that four hydrogen nuclei—four protons—end up as a helium nucleus of two protons and two neutrons. Neutrons are particles that are electrically neutral and a teeny bit heavier than protons. How two of the protons are converted to neutrons is a question that need not concern us here; the answer is known, if you want to look it up in a nuclear physics text.

In an earthly laboratory, where physicists must use ingenuity instead of gigatons of matter to induce fusion, the scientists work largely with an isotope of hydrogen called deuterium. An isotope is a variation of an element in which the atomic nucleus has a few more neutrons (or a few less) than the "normal" element does. The resulting atom is the same chemically, and therefore considered to be the same element. But its nucleus is somewhat heavier (or lighter) than the "normal" nucleus.

The deuterium nucleus has one proton and one neutron. When deuterium replaces "normal" hydrogen in water, we get what is called heavy water. It is chemically exactly the same as water, and you can drink it by the gallon to no ill effect whatever—except that it tends to disturb the balance mechanism of the inner ear, undoubtedly because it is heavier than normal water. Drinking heavy water will not harm you, but it will make you lurch and stagger as if you were drunk.

Deuterium exists naturally on Earth. There is one atom of deuterium mixed in with roughly every 6,000 atoms of ordinary hydrogen in the waters of our planet. That means that there is enough deuterium in a glass of water to equal the energy content of 500,000 barrels of petroleum. Using only a minuscule fraction of the water!

Thermonuclear fusion is so powerful that it staggers the imagination. One cubic mile of water can provide as much energy as all the known oil deposits on Earth. Since there is considerably more than 320 *million* cubic miles of seawater on Earth, the fusion process can provide all the energy we need for millennia into the future.

To put it on a more human scale, every time you flush a toilet you are pouring away about 15,000 kilowatt hours of potential fusion energy.

That is fusion's potential. Yet we must recall the old Hungarian recipe for an omelet that starts, "First, steal some eggs. . ."

No controlled thermonuclear fusion reactor exists on Earth. Despite the best efforts of the best minds over nearly half a century, the goal of fusion power still seems decades away. However, laser energy may be the key to making fusion practical.

Since the 1940s physicists have built experimental fusion reactors in which they have used powerful magnetic fields to heat small amounts of deuterium fuel up to temperatures of hundreds of millions of degrees. Under such conditions the deuterium nuclei are completely stripped of their orbital electrons; the gas consists of swarms of free electrons and bare nuclei, which are called ions. Physicists call such an ionized gas a *plasma*.

Always, the star-hot deuterium plasma manages to slip through its confining magnetic fields and cool off before any significant amounts of energy can be extracted.

An alternative method of inducing fusion is to put the fusion fuel into a microscopic lexan pellet and then zap it with laser energy. At the research laboratory of KMS Fusion, Inc. in Ann Arbor, a set of very energetic pulsed neodymium lasers successfully triggered fusion reactions in 1974.

That is still a long way from a practical fusion power plant. The laser-fusion process works by bombarding a tiny pellet of deuterium and tritium (a still-heavier isotope of hydrogen) with an extremely energetic pulse of light. The light must hit the pellet from all directions, all at the same instant. When it does it blasts the pellet and causes an *implosion* that momentarily raises the deuterium-tritium mixture to temperatures and pressures high enough to induce fusion.

For a nanosecond or so, a little artificial star glows.

Industry, government, and university laboratories are pursuing laser fusion. Most notable among the research institutions are KMS, Rochester University, the Lawrence Livermore National Laboratory in California, and the Los Alamos National Laboratory in New Mexico.

Livermore's first laser system was called Shiva, after the Hindu god. Its 20-arm laser system split a pulse from a single laser into 20 separate beams, amplified them and aligned them precisely, and delivered them to a pellet of deuterium-tritium fuel—all in about one-billionth of a second.

Shiva's neodymium-glass laser produced a pulse of 30 trillion watts, or 30 terawatts. But that pulse lasted only two-tenths of a billionth of a second. The actual power of the laser amounted to only about four ten-thousandths of a kilowatt-hour, enough energy to light a 60-watt bulb for four minutes. Not much power, unless and until it is concentrated in a nanosecond pulse.

Livermore has now built the ten-armed Nova system, which is 20 to 30 times more powerful than Shiva. At Los Alamos researchers have been using carbon dioxide lasers in systems called Helios and Antares. These also use multiple arms that bring the laser energy to bear on the target pellet from all sides at once. Carbon dioxide lasers are inherently more powerful than solid-state lasers, but they cannot produce the ultrashort nanosecond pulses that the neodymium systems do.

Laser fusion research is also being pursued in many other countries, particularly in the Soviet Union, which has the most extensive overall fusion research program in the world.

Laser fusion can only produce short pulses of fusion energy and may not turn out to be a practical system for generating power, in the end. But it does produce fusion reactions and may help physicists to learn how to produce and handle the star-hot plasmas necessary for steady-state controlled thermonuclear reactors.

A fusion generator could be adapted for space propulsion, and physicists such as Robert W. Bussard (formerly of Los Alamos National Laboratory) have suggested using fusion-based engines to power ships on interstellar journeys.

Lasers can also be used directly for space propulsion. Riding a beam of laser light may become the most efficient and economical way to fly into space.

Today's rockets use the chemical energy of propellants such as hydrogen and oxygen. The propellants are burned, and the resulting hot gas streams through the rocket nozzles and provides thrust in accord with Newton's third law of motion (every action in one direction produces a reaction in the opposite direction).

Because they must carry all the tons of propellant necessary to lift off the ground, together with the huge tanks needed to carry the propellant, rocket boosters normally are nearly 98 percent propellant and tankage, seldom more than 2 percent payload. NASA's space shuttle, for example, weighs 2,200 tons at lift-off; it carries 32.5 tons of payload maximum.

In a laser-propelled spacecraft, the laser itself remains on the ground. The laser beam reaches out to heat propellants carried by the spacecraft to temperatures far higher than could be achieved by chemical burning. The higher the propellant's temperature, the faster it flows through the rocket nozzles. The faster it flows, the more thrust it produces.

The ideal laser-driven rocket would carry hydrogen, since it is the lightest element and therefore would produce the highest flow velocity per watt of laser power input. The laser beam might be fired directly into the throat of the rocket nozzle itself, or it might be aimed at a special receiver built into the spacecraft and channeled into the rocket with mirrors.

Such a spacecraft could be 50 percent payload. The laser beam provides the energy, transferring it from the ground to the ascending rocket booster with the speed of light. The ground-based laser can be used over and over again.

Laser propulsion studies are under way at Physical Sciences Inc. in Massachusetts, at several universities, and at NASA's Marshall Space Flight Center in Alabama.

Robert L. Forward, a physicist at Hughes Research Laboratories in Malibu (and a science-fiction writer as well), foresees huge lasers in orbit around the Earth that propel spacecraft across the Solar System and even farther, out toward the stars, on their powerful beams of light.

Forward champions the idea of laser-propelled *lightsails,* enormous gossamer kites many miles wide that are pushed through the vacuum of interstellar space by the pressure of laser light. It is well known that light can exert a minuscule amount of pressure, and "solar sails" will be tested within the next decade for flights between the Earth and Moon, propelled by the pressure of sunlight. Forward wants to go a step beyond that; instead of using sunlight, he wants to use the more concentrated pressure of laser beams to propel lightsails out to the stars.

That old Hungarian recipe for an omelet confronts us once again. Before you can have laser-propelled spacecraft or starspanning lightsails, you need incredibly powerful lasers. While several research teams are experimenting with toy-sized models to work out the mechanics of laser propulsion, the lasers they are using are thousands of times too weak to be of practical use.

It may be, however, that the lasers being developed for the

Strategic Defense Initiative could be powerful enough to be practical for propulsion. If "Star Wars" weaponry results in laser-propelled starships, we will see a new example of swords being beaten into plowshares.

The world is filled with coincidences, some of them bitterly ironic, others beautifully appropriate. What can be more lovely than using lasers—the most sophisticated form of light that human ingenuity has yet produced—to restore sight, to help develop artificial stars that will light our world for millions of years to come, and to propel human explorers to the stars themselves?

Extremely powerful lasers in orbit may one day power lightsails, propelled by the pressure of the laser's light, beyond the edges of the Solar System and out toward the stars.

17

Fibers, Disks, and Solar Cells

SCIENCE FICTION can roam the distant past as well as the far future. In a novel of mine titled *Vengeance of Orion* I set part of the action in ancient Egypt. My hero, Orion, is surprised to find that news of happenings in the Nile delta gets to the capital city, hundreds of miles up the Nile, in less than a day.

> I blinked with surprise. "How did you get a message. . ."
> Nefertu laughed, a gentle, truly pleased laughter. "Orion, we worship Amon above all gods, the glorious sun himself. He speeds our messages along the length and breadth of our land—on mirrors that catch his light."
> A solar telegraph. I laughed too. How obvious, once explained. Messages could flash up and down the Nile with the speed of light, almost.

Jump forward 3,200 years.

"The optics [communications] boom is just starting to explode," says Robert Spinrad, director of systems technology at Xerox Corporation. "Optics in the twenty-first century will be what electronics represents in the twentieth century."

The use of light in modern communications is based on two inventions. One of them is the laser, of course. The other is fiber optics, strands of ultrapure glass thinner than a human hair, capable of carrying laser light for miles. Engineers now call them *light pipes* or *lightguides.*

While earlier forms of light communications depended on sunlight and were useless at night or in bad weather, fiber optics represents the fastest-growing technology in the communications industry.

Light has been used to carry simple messages for thousands of years. Mirrors sent coded signals. Bonfires were used as signals. Paul Revere watched on the north bank of the Charles River for signal lamps in the tower of Boston's Old North Church: one lamp if the Redcoats were going to head for Lexington and Concord by marching overland, two lamps if they were going to cross the river by boat and then march northward.

In a sense, semaphore signals and even hand gestures are light-based communications. Some languages (Italian, I think, in particular) lose half their meaning if the spoken words are not accompanied by the appropriate dramatic gestures. But here we shall consider only communications systems that can carry more complex information than a semaphore flag or a wagging finger.

Alexander Graham Bell, inventor of the telephone, also invented the Photophone, a device that transmitted voice signals over a beam of reflected sunlight. Bell actually considered the Photophone the more important invention of the two, reasoning that sunlight would be more reliable and far cheaper a communications link than electrical wire. He was so excited about the Photophone that he wanted to name his second daughter after it, since she was born a few days after his first successful demonstration of the device, in February 1880. Apparently Mrs. Bell did not share his enthusiasm.

The Photophone worked by using the sound waves of the speaker's voice to vibrate a mirror, which then sent flickering pulses of reflected sunlight to the receiving instrument. The receiver had a rod of selenium, a metal whose electrical resistance changes with the intensity of light falling on it. When the selenium rod was connected to a phone receiver and a battery, it faithfully reproduced the speech sent from the transmitter.

Electricity turned out to be far more reliable than sunlight, however, and telephone cables began "decorating" city streets and rural vistas all over the globe.

Meanwhile (in fact, ten years earlier), unnoticed by most of the world, a British physicist laid the basic groundwork for optical communications. He showed how to bend light around a corner.

John Tyndal shone a light into a spout of water as it gushed out of a tank. The water fell in an arc toward the ground, and the light went with it, following the same curve. It was as if the light, once inside the water spout, was trapped there. The water acted like a light pipe, which indeed it was.

What was happening was that the outer edge of the water spout acted like a mirror, reflecting the light that reached it back toward the interior of the water spout. This is called *total internal reflection*. When light that is traveling through a dense medium, such as water, hits the boundary between that medium and a medium that is less dense, such as air, it is totally reflected back inside the dense medium. The air-water boundary acts as a mirror. One caveat: Total internal reflection works only when the light strikes the boundary at a small, glancing angle. At steeper angles the light passes through the boundary just as if it were a window, rather than a mirror. A light pipe will turn light around a corner only if the turn is gentle, gradual. Sharp angles are out.

In 1880, the same year that Bell got so excited about his Photophone, William Wheeler of Concord, Massachusetts, applied for a patent on a light pipe. His idea was to use pipes whose inner surfaces were highly reflecting to distribute light from a central bright source to other points in a building—sort of like the way heating ducts distribute heat from a central furnace to all the rooms of a house. Wheeler got his patent, but his light pipes never worked very well. He planned to use silvered mirror surfaces inside the pipes, and they simply absorb too much light to serve

Optical fibers (sometimes called light pipes) work on the principle of total internal reflection. *Light striking the boundary between the core and the outer cladding is reflected along the core. Light striking the boundary at a steeper angle passes through the cladding and is lost.*

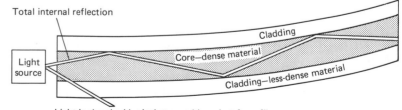

Total internal reflection

Cladding

Core—dense material

Light source

Cladding—less-dense material

Light is absorbed by jacket on cable or lost from fiber

as effective reflectors for a light pipe. The light did not get very far before the mirrors absorbed it all.

Enter a latter-day Robin Hood in the form of another English physicist, Charles Vernon Boys. Not that he stole from the rich and gave to the poor, but he was something of an archer. Boys produced the first glass fibers by attaching molten quartz to an arrow and firing it off from a bow. This yielded fibers that were hair-fine yet strong enough to be used for hanging delicate physics apparatuses. Boys shot an arrow into the air and made a fiber thin as hair. (With apologies to Longfellow!)

The idea of using light to carry communications transmissions, the concept of total internal reflection for bending light around corners, and the ability to make fine glass fibers were all in place by 1880.

Move forward 54 years.

Norman R. French of AT&T obtains a patent for an "optical telephone system." His idea is to send voice signals over a beam of light that is carried through "light cables." Brilliant idea, but the technology for it does not yet exist.

Jump forward another 26 years.

The laser is invented. Two years later, in 1962, three separate research teams—at IBM, at MIT's Lincoln Laboratory, and at the General Electric Research Laboratory in Schenectady, New York—announce the invention of the semiconductor, or diode, laser. Good things come in small packages, it is said. Also in threes.

Researchers at AT&T's Bell Labs, realizing the enormous communications possibilities of the laser, examined the idea of transmitting telephone and video signals by laser beams. As we have seen, light waves offer millions of times more information-carrying capacity than radio waves or electrical signals routed through copper wires.

But transmitting laser beams through the atmosphere brings on all the problems that defeated Bell's original Photophone. Rain, snow, fog, pollution, and even normal humidity absorb or distort laser beams. What was needed was a light pipe.

Optical fibers had been available since the 1950s. They were developed by Brian O'Brien, Sr., at the American Optical Company in the United States and, in Britain, by Narinder S. Kapany and his colleagues. However, these fibers could not carry a light signal more than a few yards.

By the mid-1960s researchers at the British Post Office, which operated the British telephone system, began to look into the possibilities of using fiber optics as light pipes. Charles K. Kao and George A. Hockman, engineers at Standard Communications Laboratories (a British subsidiary of ITT), predicted that glass fibers could be made so pure that light could be transmitted for at least a mile through them.

Robert D. Maurer of the Corning Glass Works heard about this work during a visit to Britain. He went back to Corning's laboratories in upstate New York and by 1970 had produced the first practical fiber optic lightguide.

Maurer was a physicist who had not been involved in fiber optics until then. He believes he and his Corning colleagues were successful because they approached the problem "with fresh eyes."

The problem was to make optical fibers that could carry laser beams for distances of a mile or more. Earlier approaches generally tried to use compound glasses: a dense glass core for transmitting the light and a less-dense "cladding" around this core that would reflect light inward, like the air-water boundary of Tyndall's experiment.

Working with fused silica, the Corning group produced the first practical optical fiber by 1970. It could transmit light about a third of a mile before 90 percent of the light was absorbed. Within ten years fiber optic cables were transmitting light 20 miles. In a test in 1983, Bell Labs transmitted an optical signal through a 100-mile length of optical fiber without a repeater to boost the signal. At the time this is written, that is the world record for optical transmission. But the record probably will not last long.

Fiber optic lightguides are now so transparent that if seawater were as clear, you would be able to see the bottom of the deepest ocean trenches.

The technology for light communications was finally in place, a century after Bell's Photophone:

1. Semiconductor lasers, small as transistors, that emit coherent light capable of carrying more than a billion data bits per second;

2. Fiber optic light pipes, or lightguides, to carry the signals for distances of miles; and

3. Repeaters to boost the light signals before they become too faint. The repeaters are light-emitting diodes, similar to the LEDs of digital clocks, pocket calculators, and other electronic appliances. With efficient repeaters no bigger than the diode lasers themselves, light signals can be sent across thousands of miles.

Copper telephone cable, three inches wide, consists of 1,500 pairs of wires. It can carry 20,000 two-way voice signals. It needs

Optical fibers transmit laser light that carries thousands of telephone conversations. If seawater were as clear as the glass of these optical fibers, the bottom of the deepest ocean trenches would be clearly visible.

electrical repeaters to boost the signal every mile, and it costs roughly $30 per yard.

The core diameter of a fused silica fiber is typically eight microns, about three ten-thousandths of an inch—actually ten times thinner than human hair! The outside diameter is 125 microns. Fiber optic cables are half-inch-wide bundles of 144 of these ultrathin fibers. They can carry 80,000 two-way signals today and will handle 240,000 signals by the 1990s. Optical fibers cost about twice as much as copper cable at present, but they carry four times the workload and will soon be carrying 12 times the load.

Throughout the world, fiber optic lightguides are replacing the copper wires of telephone systems. In 1984 AT&T announced a lightguide fiber that can carry 200 million data bits (two megabits) per second. The following year Bell Labs demonstrated the feasibility of sending 300,000 telephone calls or 200 high-resolution TV signals over a single optical fiber. By 1985 researchers had transmitted four gigabits per second through a lightguide cable, the equivalent of 62,500 simultaneous telephone conversations.

Transatlantic cables have carried telegraph and, later, telephone communications since 1866. The eighth transatlantic cable, TAT-8, is the first to use fiber optics. It can carry 37,800 simultaneous telephone calls or a mix of voice, television, and computer data signals, four times the capacity of the copper coaxial cable TAT-7. TAT-9 is already being planned by AT&T and the telephone organizations of Canada, Britain, and France. It will use light-wave technology and be capable of carrying 1,130 megabits per second, twice the capacity of TAT-8.

Laser signals can be coded to carry *digital* data. The world's telephone systems are evolving into a giant interconnected digital computer, and lasers with lightguides are making it increasingly a computer with optical interconnections.

When you speak into a telephone, your voice is transformed inside the phone into an electrical current. The current goes to a telephone exchange, where the message is routed to your intended listener. The telephone converts the fluctuations in air pressure caused by your voice into fluctuations of the electric current, then converts the current back to audible sound at the receiving end of the conversation. This is called an *analog* system.

At more and more telephone exchanges, though, the incoming electrical system is converted into digital bits so that it can be processed by the computers that handle the task of routing your call swiftly and correctly. Voice signals are carried at a rate of 64,000 bits (64 kilobits) per second over copper wires, while lightguide fibers routinely carry 90 megabits and, as we have seen, four gigabits have been transmitted experimentally.

Digital systems are not only powerful, they are also flexible. Digital phone systems can handle voice, computer data, and video signals with equal ease. Performance quality is superior to analog transmission because the digital signals are less likely to be distorted in transmission than a continuously varying electrical current.

In the United States most urban telephone exchanges are now digital, and pulses of laser light are rapidly replacing electrical current. "Optical signal processing is the wave of the future," says Ira Jacobs, director of Bell Labs' wideband transmission facilities laboratory.

Fiber optics and lightguide communications are already changing the world. Long-distance fiber links are challenging communications satellites as the best and least expensive way to send voice and TV transmissions across continents and oceans. Moreover, a long-distance telephone conversation transmitted through a lightguide fiber does not have that annoying half-second lag that communications satellites force upon us. Even at its phenomenal speed, light (in the form of microwaves) takes a noticeable fraction of a second to go from one phone to a commsat 22,300 miles above the equator and then down again to the receiving phone.

Optical communications and optical computers, based on the immense information-handling capacities of lasers, will lead to the marriage of telephone, computer, and TV set. The resulting device will be a communicator that can tap into the Library of Congress as easily as it can obtain the latest weather forecast or connect you with a friend halfway across the world for a face-to-face video conversation.

Ultimately the communicator will be small enough to carry around with you, small enough even to wear on your wrist.

Fiber optics have allowed physicians and surgeons to truly see into the human body. A hair-thin lightguide can carry light into the deepest crevices of the body's interior and allow physicians to actually see what is going on inside, in real time, without the need for exploratory surgery. When surgery *is* called for, lightguides can carry pulses of laser light precisely to the spot where it is needed. Tumors in the brain, bladder and other locations that could not have been reached by the surgeon's knife without doing irreparable harm to the patient are now excised cleanly and neatly by purifying laser light carried deep inside the body on fiber-optic lightguides.

Fiber optics systems will begin to replace electronic and electrical controls in the new aircraft being developed for the 1990s. The X-30 National Aerospace Plane, which will be capable of flying into orbit from ordinary airfields, will use beams of light guided through optical fibers to flash information through its control systems.

While fiber optics can transmit information by light waves and carry surgical laser beams into the body's organs, laser light can be used to *store* information as well. We saw in the previous chapter the possibilities of photonic computers with holographic memories. Such possibilities are for the future, perhaps the relatively near future, but they are not yet with us.

However, optical information storage systems are among us in our everyday lives. Compact disk (CD) records are made, and then played, with laser beams. So are videodisks. And optical disks are being developed to store data efficiently and compactly for hospitals, businesses, and other organizations that generate huge quantities of files.

Laserdisc, videodisk, optical disk—the names (and spellings) vary, but the principles are the same. Laser beams put information onto a disk, and laser beams are used to read the information from the disk.

The key to this technique is the fact that a laser beam can be focused down to an extremely small spot on the order of one micrometer (about four-millionths of an inch) in diameter. A light-sensitive film is coated onto a disk, and a laser burns tiny holes into the film while the disk is rapidly spun. The laser is operated in extremely short pulses, so that each spot on the disk is kept as small as possible.

The holes are burned onto the disk in a pattern, like a sort of microminiaturized braille, that can be read by another laser. The "reading" laser is of lower power, so that it does not burn more holes into the disk. Its task is to "see" where the holes are and report that information to the system's electronic equipment, which interprets the pattern of holes into digital sound, video pictures, or other forms of information display.

Compact disk audio records are the most common form of laser disk in use today. A 4.75-inch CD contains more than three miles of digital laser-burned holes on each side. When you play the disk, nothing touches it but the laser's "light" touch, which means that the disk will never wear out. Also, the music coming from the disk has none of the crackle or hiss of needle-played records; its digital sound is the best audio reproduction ever.

In the late 1970s several large corporations such as RCA, Philips, and Japan's JVC introduced videodisks—laser disks that recorded motion pictures for viewing on home television sets. The technology was fine, but the product was a colossal disaster in the marketplace. Consumers preferred videotape far better than videodisks, mainly because videotapes can be erased and reused, while videodisks are permanent.

One of the drawbacks of laser disk information storage is that once those holes have been punched in the disk's optically sensitive film coating, it is extremely difficult (for the home consumer, impossible) to erase them and rerecord something else on the disk.

However, the capability of storing some 10 billion bits (10 gigabits) of information on a disk less than five inches across, and accessing the information in microseconds, can't be all bad. Predictions from the business community claim that optical data storage for computers will become a multibillion-dollar business within a decade.

For example, Optical File Management, a new Massachusetts firm, is developing data systems in which files and other forms of archival information can be stored and reproduced optically. A document can be retrieved from the system to be viewed on a display screen or photocopied. Paul Kellen, founder of the company, stresses that such systems store *images* rather than data. Instead of printing out numbers or letters, the system reproduces the original piece of paper. This offers more reliability and efficiency, especially when the originals were handwritten—such as day-to-day medical records in a hospital.

There is another way that we use light, a technique for generating electrical power directly from sunlight, employing *solar cells*. While solar cells are used almost entirely in outer space at present, they are beginning to find applications here on the ground. They have the potential of making the greatest change in our lives since Edison's day.

Sunlight is energy, electromagnetic energy, pure and simple. Certain materials called semiconductors can transform sunlight directly into an electrical current. Semiconductors are the basis for all the transistor chips that run our computers and other solid-state electronic devices. Solar cells are essentially semiconductors that make use of the photoelectric effect: Photons of sunlight strike the semiconductor material and produce a flow of electrons—an electric current.

The earliest use of solar cells has been to provide electrical

Solar batteries, or "cells," began by powering artificial satellites and are now used in pocket calculators and many other common electrical appliances.

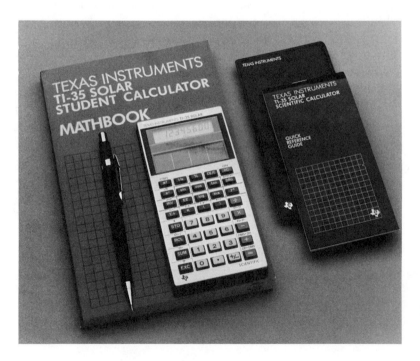

power for spacecraft. The first *Vanguard* satellite, launched in 1958, beeped away for more than a dozen years thanks to a few tiny solar cells that transformed sunlight into electricity. The Soviet space station *Mir*, in orbit since early 1986, generates enough electricity to take care of six or more cosmonauts and all their equipment by using broad panels of solar cells.

Here on Earth solar cells are used in special applications. Some roadside telephones in remote areas of the southwestern desert region of the United States are powered by solar cells. Sunlight is plentiful, and it is cheaper to provide the phones with their own power source than to string cables for many miles.

In a recent Christmas shopping catalog I saw an advertisement for solar-powered lights that can be placed along your driveway or the walk in front of your house. They have a battery in them that stores the electricity generated during daylight and then powers the light for the first few hours of darkness.

Solar cells are still too inefficient for most commonplace uses. Typical efficiencies run slightly better than 10 percent. While this is okay for certain applications such as remote telephones in the desert, their low efficiency means that solar cells are too expensive for household use. But this situation will undoubtedly improve in the foreseeable future, and solar cells may become cheap enough for everyone.

Remember that there are dozens of kilowatts of solar energy falling on your roof every moment of the day. If solar cells could be made 40 or 50 percent efficient, it would become possible to generate all the electricity your household requires out of sunlight. You could disconnect from the electric utility company and become energy-independent.

In fact, recent work done in Israel has resulted in a solar cell that is also a storage battery. Not only can it generate electricity directly from sunlight, but it can store electrical energy for use at night. The research was reported in 1987 by physicist Stuart Licht, of the Weizmann Institute of Science. The word *licht* comes from the same Germanic root as our word *light:* another delightful etymological coincidence.

While it may be possible eventually to make your home energy-independent, don't think that the electric utility companies are going to disappear. Cities and factories are such intense users of electrical energy that rooftop solar cells may not be able to provide all the power they need.

But solar cells in space may.

Picture a satellite hovering 22,300 miles above the equator, out where communications satellites now orbit. This is no commsat, however. This object is immense, 12 miles long: the size of Manhattan Island. It consists almost entirely of solar cells. It is a *solar power satellite.*

The solar power satellite (SPS) is the brainchild of Peter E. Glaser, a researcher at the Arthur D. Little Company of Cambridge, Massachusetts. Glaser's reasoning is beautifully simple and direct. It is difficult to use solar energy on Earth, Glaser figured, because weather and night make sunlight relatively unreliable. It is possible, however, to put a satellite into an orbit where it will be in sunlight all the time. Why not generate electricity from sunlight in space and transfer that energy down to Earth?

The SPS concept, therefore, envisions huge satellites that can generate solar electricity and send it to Earth through microwave beams. The energy beams would be received at special antenna "farms" on Earth, where they would be converted back into electricity. A single SPS might generate from 5 to 10 million kilowatts—5 to 10 gigawatts—enough to provide all the electrical power for an industrialized state such as Connecticut.

The SPS concept is elegant. It provides intense sources of electrical power, the kind of multigigawatt source that cities and factories need. And it does so without burning an ounce of fuel, without any pollution, because the "power plant" is 93 million miles away.

Some critics have voiced worries about beaming gigawatts of microwave power through the atmosphere. Glaser points out that the microwave frequencies used would *not* be the same as those used in microwave ovens, and the beam would be so diffuse that birds could fly through it without harm. Still, no one knows what the long-term effects of many such beams would be on the environment.

Yet we may see such immense satellites beaming energy to sprawling antenna farms in remote areas, providing such abundant electricity that today's power-generation plants can be closed

down. Acid rain would end. Fears of nuclear accidents would disappear. The greenhouse effect caused by burning fossil fuels would lessen.

And the human race will have learned how to use sunlight in a new way.

Artist's conception of a solar power satellite, which could generate thousands of megawatts of electrical power from sunlight and beam it to Earth.

18

Bright
Shining Beads

NOT ALL the uses to which we have put light deal with high technology. Light, in its many aspects, can be used to enhance a person's physical appearance.

In ancient times—in fact, long before history began to be recorded—women bedecked themselves with jewelry and augmented their natural beauty with cosmetics. So have men over the ages. Jewelry, cosmetics, and fancy clothes have never been solely the prerogatives of women.

The use of cosmetics may have started for practical reasons. In the pitiless hot sun of Egypt, before the first pharaohs, both men and women applied green coloring around their eyes. This may have originally been done to cut down on the Sun's glare, just as athletes today smear dark eye shadow on their cheeks.

By the time the first pyramids were being built, gray eye shadow was being used, together with red ocher as rouge to add color to the cheeks. Henna was used at first to bleach the palms and fingernails; only later did it serve to bleach the hair.

An equally ancient civilization arose in the Sumerian plain between the Tigris and Euphrates rivers in the land now called Iraq. But while Egypt remained relatively stable over those early millennia, the Sumerian plain saw constant warfare and the rise and fall of many kingdoms and empires. Men wore beards there,

All societies use cosmetics and jewelry to enhance beauty.

unlike the clean-shaven Egyptians, and devoted considerable care to oiling and curling their beards and hair. Men and women both used black dyes for their hair and eyebrows and henna to bleach the hair an orange-red color. On holidays, hair and beard might be adorned with gold thread, scented yellow starch, or even sprinkled with gold dust.

Perfumes and oils have been used as far back in time as archaeologists can trace and probably even further. By Roman

Cosmetics may have begun in pre-dynastic Egypt as a means to protect the eyes against the glare of harsh sunlight, just as modern athletes do.

times women were using *kohl* as a dark eye shadow, various powders to whiten their skin, and rouges for their cheeks. In Elizabethan England, women rouged their cheeks with cochineal, a red pigment obtained from dried insects.

Modern cosmetics make full use of the organic chemistry laboratory. Starting with ingredients as diverse as wool grease (lanolin), beeswax, castor oil, liquid paraffin, talc, chalk, titanium and iron oxides, carbon black, nitrocellulose resin, various alcohols, dyes, and water, the cosmetics industry produces a bewilderingly broad range of products for skin, hair, nails, lips, cheeks, and eyes—including eyelashes and brows.

Packaging is equally important, for these products could not have found wide market applications without the collapsible tube, the aerosol container, and various paper and plastic wrapping materials.

In her best-selling book *Color Me Beautiful,* beauty consultant Carole Jackson stresses that a woman's skin tone and hair color determine the kind of makeup and clothing colors she should wear. Dividing all of womankind into four groups, which she names after the four seasons, Jackson has created a color wheel reminiscent of Newton's original. However, the beauty consultant's color wheel deals with shades of unsaturated colors, not primaries and their secondaries. Newton might have been nonplussed by colors named clear salmon, bittersweet red, pastel aqua, or icy violet.

However, Jackson's major point is that the eye's natural aptitude for handling color contrast determines the beauty seen by the beholder. Some color combinations are pleasing; others are not. Some shades of lipstick, for example, can make the teeth seem glowing bright, while a different shade will make the same teeth look yellowish or gray.

The modern cosmetics industry has made it easy for people to change the color of their hair or even their eyes, with tinted contact lenses.

In addition to cosmetics, men and women have been adorning themselves with jewelry. In some cultures it has been fashionable to pierce the ears, or even the nose and other parts of the face and body, so that jewelry can be placed there. In some cultures women stretch their necks with metal bands, and the more bands and longer the neck, the more beautiful the woman is considered

Skin tone and hair coloring determine the hues of clothing and cosmetics a person should wear.

to be. In virtually all cultures, ornaments of precious metals and gems are regarded as symbols of wealth or affection—sometimes both.

Gold was among the first of the precious metals to be used for adornment. So rare that it was always highly prized, gleaming gold is also beautiful to the eye and soft enough so that it could be worked even with the simple tools of early civilizations. The metal is also "incorruptible," to use Homer's word for it; that is, gold will not tarnish or rust.

Most of the gold used in jewelry is alloyed with harder metals such as silver, copper, zinc, or nickel. Gold's purity is measured in *karats*. The karat is a percentage, a ratio, not a measure of weight or any physical quality; 24 karats equals 100 percent. Thus

24-karat gold is unalloyed, pure; 14-karat gold is 14/24ths, or about 58 percent, gold. Some jewelry is merely coated with a thin layer of gold and is called, inversely, gold filled.

Silver, copper, even bronze have been used for adornment over the millennia. Platinum, discovered in Colombia in the eighteenth century, was not used in jewelry until the first decade of the twentieth. Its major uses are still in industry, rather than jewelry.

But diamonds, as they say, are a girl's best friend.

Gemstones have a romance about them. Rubies, emeralds, diamonds, and sapphires are the stuff of buried treasure. Opals, pearls, jade fill the mind with exotic visions of *A Thousand and One Nights* or the splendid costumes of royal courts.

Think of the way a diamond sparkles. Or of how a pearl seems to glow from within. Or of the shifting, almost hypnotic, way an opal shimmers in the light.

Gemstones reflect light in ways that dazzle our eyes. There are other minerals that are rarer, other metals costlier, than gold and silver. It is the way these precious metals and glittering stones reflect light that make them so beautiful to our eyes. (See Plate 22 following page 238.)

It is not their chemical composition that makes gemstones so treasured. They are made of the most common elements on Earth, for the most part: the carbon of coal, the silicon of sand, oxygen, and aluminum. Diamonds are pure carbon, crystalline in atomic structure. Emeralds are actually a form of the mineral beryl, composed of beryllium, aluminum, silicon, and oxygen. Rubies and sapphires are forms of corundum, which is composed mostly of aluminum and oxygen.

What makes them so valuable? To begin with, a gemstone must be beautiful. It must glow with inner fire, catching the light so that it almost seems alive with color. Gemologists speak of iridescence, opalescence, asterism, luster—they are speaking of the ways in which gems reflect light.

A gemstone must also be rare. There is plenty of beach sand, carbon, aluminum, and oxygen in the world. There are precious few diamonds, emeralds, and rubies. And how many oysters have you eaten in your lifetime? Have you ever found a pearl?

Gemstones must also be durable, especially if they are to survive the cutting and polishing they must face before they can be offered as jewelry. Some are more durable than others. Diamond is one of the hardest substances known. Opals are relatively soft and rather easily broken.

Gems are measured by weight. The unit used is the *carat*, which is 200 milligrams, or approximately seven-thousandths of an ounce. Do not confuse carats with karats. Gemstone carats can be measured on a balance. The famous Hope diamond is 44.5 carats. The Star of Asia is a sapphire of 330 carats, or about 2.3 ounces. The Star of India is a slightly oval gray-blue sapphire that is polished but unfaceted, in the cabochon style of cut (see below); it weighs 536 carats, or 3.752 ounces.

Long ages ago it was discovered that a gem's beauty could be enhanced (and minor flaws in the stone hidden) by grinding or cutting flat surfaces onto the body of the stone. Today gemstones are cut to bring out their color or brilliance. While in the Orient the weight of the stone is regarded as paramount, so that as little cutting is done as possible, Westerners value the overall beauty of color and symmetry more than weight alone and do not mind if a fair amount of gem is sacrificed by cutting. In the West, therefore, gems are cut to enhance their light-catching ability. Usually they are cut into many facets, each facet following a natural plane of cleavage in the stone's crystalline structure. Naturally, gem cutting evolved as more of an art than a science. Even today, the gem cutter is a highly regarded craftsman; when he makes a mistake, a precious stone can be ruined or even shattered completely.

The major kinds of cuts are as follows.

1. *Cabochon,* with a flat circular or elliptical surface atop a dome that may be shallow or steep, depending on the original stone. Used most often for opaque gems such as turquoise, iridescents such as opal, and opalescents such as moonstone.

2. *Rose,* which often consists of six triangular shallow facets on top, surrounded by 12 steeper triangular facets. Often used for transparent stones such as diamonds.

3. *Brilliant,* originated in seventeenth-century Venice, is the first cut to take advantage of the natural refraction angles in a diamond. The cutter calculates the angles between facets so that light entering the diamond from the side facets

is totally reflected through the topmost facet (the crown) into the eye. The farther the light must travel within the diamond, the more it will be refracted, spread out into a fiery rainbow of colors, when it finally emerges. Thus the brilliant cut is made with as many facets as possible to force the light through a veritable "hall of mirrors" inside the stone before it emerges through the crown. Vincente Peruzzi, credited with inventing the brilliant cut, put 24 facets on the "pavilion," or lower portion of the diamond, plus 32 facets on the crown, surmounted by a flat "table" through which the light finally emerged.

The brilliant cut is used for other gems in addition to the diamond, but the angles of the facets are always determined by the individual stone's index of refraction.

4. *Step or trap* is a cut similar to the brilliant, in that there is a lower-section pavilion topped by a crown. The basic shape is square or rectangular in cross section. This is also known as the *emerald cut*.

There is also the *mixed cut,* usually with a brilliant-cut crown and a step-cut pavilion.

In the May 1988 issue of *Smithsonian* magazine, writer P. F. Kluge reports on the three-year project to convert an 890-carat diamond, the fourth largest ever found, into a 531-carat finished gem. The color photographs show every step along the way from raw stone to gleaming "triolette."

Precious metals and gems have many uses in industry. Platinum, for example, is an excellent chemical catalyst; you tote around several hundred dollars worth of it in the antipollution catalytic converter built into the exhaust system of your automobile. Gold is an excellent heat insulator; many spacecraft have been gold plated not as a symbol of extravagance but because a thin layer of gold was the most efficient way to keep the spacecraft from overheating.

Diamonds have long been used in industry for cutting and polishing. Indeed, in the world of jewelry itself, diamond dust is used to polish diamonds.

Aerospace engineers are beginning to make use of diamond's unique combination of optical, physical, and electronic properties for a wide variety of uses in aircraft and space vehicles. They have learned to make thin films of diamond by heating commonplace methane and hydrogen gas with microwaves or radio energy to a temperature of about 1,000°C. The carbon is literally boiled out of the methane at that temperature and can be deposited as pure diamond film on a substrate. This process is called diamond *vapor deposition.*

Diamond film is strong and tough. It can transmit visible, infrared, and ultraviolet light. It conducts heat five times better than copper and is one of those rare materials that is an electrical insulator yet a thermal conductor.

Diamond films are now being developed for use as semiconductors that will handle much more electrical power than silicon chips, as tough coatings for windows and sensor domes on hypersonic aircraft and space satellites, even as output windows for multimegawatt lasers.

When human explorers head out toward the stars, they will bring gold and diamond with them—not only as personal adornments but as practical and efficient materials for their starships as well.

IV. TO SEEK

I saw eternity the other night,
Like a great ring of pure and endless light,
    All calm, as it was bright;
And round beneath it, Time, in hours, days, years,
    Driv'n by the spheres,
Like a vast shadow moved, in which the world
    And all her train were hurled.

—Henry Vaughan

19

The Lights in the Sky

IT BEGAN in the desert.

The stars shine brightly in the cold dry air of the desert night, filling the sky with flickering points of light, dazzling the eye with their splendor.

Long before people learned how to write, before they began to build cities or even to plant crops, desert nomads watched the stars. Named them. Observed their motions through the crystal darkness. Wondered who had put them there and what unexplained force guided their motions.

In the clear air of the desert the stars seem very close. They swing across the night sky like a gigantic wheel, rising in the east and setting in the west. They march across the heavens rank on rank, always in the same order, eternal and unchanging.

Or so it seems.

Those early star watchers found mysteries in the night sky, mysteries that must have puzzled them from generation to generation even as they guided their steps across the trackless desert by the light of the pole star and its companions.

One of the most puzzling aspects of the night sky is that, while nearly every star remains in the same formation night after night, year after year, there are five visible stars that wander slowly across the sky, blatantly ignoring the unchanging perfection of

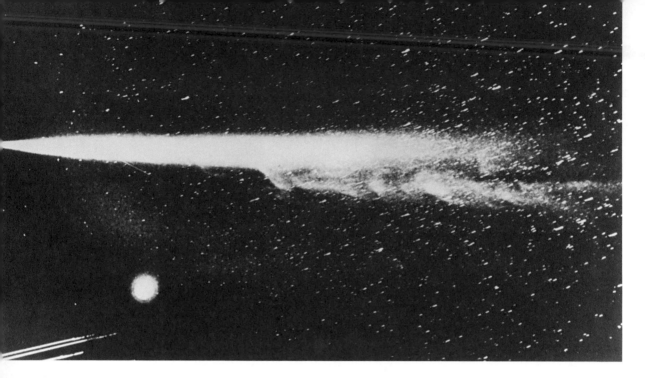

Ancient peoples thought that the pale apparition of a comet presaged catastrophe.

their fellows. If the human mind has one particular trait, it is the search for *order*. Thousands of stars move in unchanging order, generation after generation. Five do not. How so?

And sometimes a star falls, flashing briefly through the night to wink out and disappear forever. A startling thing to see. More frightening still, once in a great while there is the ghostly apparition of a comet, hanging pale and ominous like a dreadful finger pointing toward some awful disaster yet to come.

The most terrifying thing of all, though, must have been eclipses, when the Moon fades from sight or turns blood red, or the Sun itself is eaten away and darkens, its life-giving warmth ebbing away as you watch.

Clearly the heavens were the domain of forces far beyond the control of mere mortals. Clearly this was the province of mighty gods, a realm totally different from the squalid dirt of humanity's habitat. The stars were far beyond the reach of man.

But not beyond human understanding. At least, some men began to try to make order out of the magnificent panoply of the stars. They saw that the stars moved around the Earth in a regular fashion and that their motions coincided with the seasons on Earth. They found that some groups of stars seemed to be arranged in a way that reminded them of earthly creatures: In those groupings they saw pictures of a lion, a hunter, dogs, bears, kings and queens, and more. They named these *constellations* after the fanciful shapes they saw.

The five wandering stars were obviously more powerful than those that remained meekly in place, so these were named after their gods and goddesses. Since Western civilization is the descendant of the Roman Empire, today we call them by their Roman names: Mercury, Venus, Mars, Jupiter, and Saturn.

Roughly 10,000 years ago, agriculture was invented. Some human tribes gave up their nomadic life and settled down in specific spots to grow crops. They built villages and towns and, eventually, mighty cities. Our word *civilization* means, essentially, "living in cities." Some men became able to devote their entire lives to watching the heavens. Why should hard-working farmers or hard-fisted overlords allow certain men to lounge about gazing at the stars rather than working like everyone else? Obviously, stargazing was not regarded as a waste of time; it was looked upon as a necessity. (Would that our modern civilization was as wise in that regard as those early ones!)

The early agricultural civilizations tolerated stargazers for a variety of reasons, both religious and practical. A society dependent on farming requires a reliable method for predicting the seasons. Plant the crops too soon and they will be killed by late frosts; plant too late and winter will set in before the crops ripen.

The immense megaliths of Stonehenge and elsewhere in Britain and France were built to serve as astronomical computers. To this very day it is possible to use those huge stones to determine

Stonehenge and other megalithic monuments were built more than 2,000 years ago to serve as astronomical computers. Midsummer's day, the changes of the seasons, and even lunar eclipses can be predicted by using the megaliths.

midsummer's day, to predict lunar eclipses, and to calculate the change of seasons. In the cloudy damp chill of the British Isles it is not easy to know when spring arrives. Those giant stones, built some 2,000 years before Christ by men who had neither engines nor cities nor iron tools, represent a colossal effort that must have been driven by the absolute need to know when to plant. The alternative was guesswork—and starvation.

Even in the very warm and stable climate of Egypt it was necessary to predict when the Nile would flood. Astronomers learned that when the very bright star we call Sirius comes above the horizon just before sunrise, the great river will soon begin its annual flood.

It must have seemed abundantly clear to these ancient men that there were powers in the stars that affected, even controlled, events here on Earth. The sky was where the most powerful gods dwelled, so the motions of the stars were the work of those gods. If someone could interpret what the gods were trying to tell us through the motions of the stars, he could help the people live in better harmony with the universe. Thus astrology—the attempt to predict the future of individual human lives—was born hand in hand with astronomy.

(A personal opinion: Astrology was, is, and always will be entirely rubbish. The positions of the stars at the time of your birth play no discernible role in the events of your life. If you want to have some fun, gather a group of friends together and read a few of the day's astrological forecasts from the newspaper without mentioning which sign each forecast is for. Then see if your friends can figure out which forecast applies to each individual. They can't, because the forecasts are so general in nature that they can fit anyone, anytime. Astrological accuracy is in the eye of the beholder, not in the precision of the forecasting technique.)

The early astronomers worked with their eyes and a few simple instruments for measuring the angular distances between stars. They could do little except plot the positions and motions of those lights in the sky. Still, they had brains. They were as intelligent as you or I. What they lacked was knowledge, information.

They built a theoretical image of how the universe was constructed. Today we know they were off the mark in many ways, but their ideas were useful for thousands of years. If we were thrown onto the desert today, without instruments and with our knowledge of modern astronomy magically erased from our minds, we would most likely come up with a very similar view of the universe.

The earliest astronomers pictured the sky as a huge bowl hanging over a flat Earth. The stars were specks of fire peppering the surface of that bowl. The Earth was at the center of the universe, flat and unmoving. The stars wheeled around it.

What made the stars move was a question they could not answer. They attributed the motion to the "nature" of the stars— the stars wheeled around because it was in their nature to do so, just as it is in the nature of Earth to be solid and unmoving.

Most of the earliest attempts to explain the movements of the stars were steeped in religious or superstitious attitudes. The ancient Greeks, however, began to put matters on a firmer, more factual basis. They realized that those five wandering stars did not rove at random all over the heavens but stayed to certain paths. The Greeks called these five *planetos,* meaning "travelers," and found that their motions were *predictable.*

The ability to predict is crucial. It is the essence of knowledge. The practical test of knowledge is the accuracy of the predictions the knowledge allows you to make. If you cannot make an accurate prediction, you do not understand what's going on. Once astronomers could predict—even with only marginal accuracy— the positions of the planets, it took much of the mystery out of them. It allowed us to look at the heavens in a new way. The motions of the stars were not entirely due to the whim of the gods. The motions were constrained, orderly. They could be understood.

By the second century A.D. the ancient land of Egypt had long been dominated by Greek culture. No matter that the Romans held political sway; Alexandria was a Hellenic city. Claudius Ptolemaeus, known to us as Ptolemy, an Alexandrian astronomer, set down all that was known about the universe in a book.

Ptolemy's book is known to us as the *Almagest,* an Arabic title. In the violent centuries that followed the collapse of the Roman Empire, it was the Arabs who absorbed and protected the hard-won ancient knowledge of the heavens—including Ptolemy's work. While Europe sank into the Dark Age, Arab scientists and physicians treasured the knowledge they gleaned from the decaying west. Most of the stars we see at night have Arabic names: Vega, Rigel, Altair, Aldebaran, Betelgeuse, Deneb, to name only a few of the brightest.

The Persian Omar Kayyam, known to Europeans mainly as a poet, was in fact an astronomer and mathematician. He may even have guessed at the idea that the Earth revolves around the Sun three centuries before Copernicus. He hints at it in this quatrain from the *Rubaiyat:*

> For in and out, above, about, below,
> 'Tis nothing but a Magic Shadow-show
> Played in a Box whose Candle is the Sun,
> Round which we Phantom Figures come and go.

Kayyam died in disgrace and obscurity. Some historians believe he incurred the enmity of conservative Islamic religious leaders who were horrified at his astronomical ideas, which they believed contradicted the sacred teachings of the Koran.

But that happened a thousand years after Ptolemy. Ptolemy's *Almagest* summarized most of ancient man's understanding of the heavens. In beautiful detail, he described the appearances of the stars and planets and tried to explain how the universe was constructed and how it worked. Thus we today speak of the *Ptolemaic system* of the universe, even though Ptolemy did not actually invent the system. He merely put together the work of earlier astronomers and philosophers, most notably the Greek Hipparchus, who had lived two centuries earlier. But Ptolemy wrote the book that survived, and Hipparchus's ideas are known today as the Ptolemaic system—the power of the pen, and the luck of a book that survived.

The Ptolemaic system firmly subscribed to the idea that the Earth stood still and was at the absolute center of the universe. The Sun, Moon, and planets all revolved around the Earth. This is the *geocentric* point of view, and it held sway for more than a thousand years after Ptolemy. It is still used in teaching navigation!

Ptolemy, like Hipparchus before him, realized that the Earth is a sphere. In fact, in the Ptolemaic system the universe is depicted as a series of spheres. Each planet, as well as the Moon and the Sun, was believed to be attached to an invisible crystalline sphere whose rotation provided the motive force for the body's observable motion through the sky. The "fixed" stars were attached to the outermost sphere. Although these ideas are totally outdated today, when we speak of "the music of the spheres" we are paying homage to the Ptolemaic view of the universe.

The revolutionary thinker who shattered this concept of the universe was no flaming rebel throwing his new intellectual ideas into the face of a thousand years worth of common sense and established order. The man who dared to suggest that Ptolemy and everybody else were wrong was a mild-mannered Polish church-

A Dutch engraving from Renaissance times pictures the Ptolemaic description of the universe, with the Earth at its center.

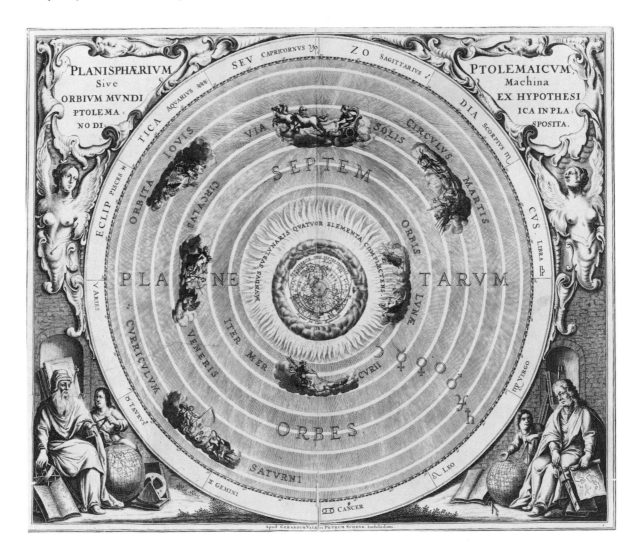

man, a man so timid that he allowed his ideas to be published only when he was on his deathbed.

His name was Nicholas Copernicus; at least, that is the anglicized version of the Latin version of his Polish name. In 1543, as he lay dying, his tearful friends showed him the first edition of his book, *De Revolutionibus Orbium Coelestium (On the Revolutions of the Celestial Spheres).*

Copernicus suggested, in *De Revolutionibus,* that it was the Sun that stood at the center of it all, and the Earth was merely one of the planets revolving around it. Learned men had known since before Ptolemy's time that the Earth was round. Columbus had discovered the New World half a century earlier. Copernicus went further, though, and removed the Earth from the center of the universe. The Copernican system was *heliocentric,* Sun-centered. The Earth moved.

He waited until he was dying before publishing his work because he understood that the Roman Catholic church would not take kindly to any new ideas that seemed to contradict the teachings of the Bible. Even so, this gentle cleric presented his heliocentric idea as a hypothesis, a figment of the mind to be pondered over, not as verifiable fact. In truth, Copernicus could not verify his conception. The proof that the Sun is at the center of the Solar System was not found for another 66 years.

However, the Copernican revolution, as it came to be called, was helped enormously by a technological innovation that had occurred almost exactly a century earlier. In 1454, in the German city of Mainz, Johann Gutenberg published the first book printed from movable metal type. His first book was a Bible, but the printing press made books of all kinds available to a wide reading audience. Instead of being rare and treasured secrets, books became common. And not all of them agreed with the Bible. Copernicus's *De Revolutionibus* was printed in Latin, as learned tomes were in that day, not to restrict its popularity but to make certain that scholars all over Europe could read it easily. Latin was (pardon the pun) the lingua franca of European scholarship.

A century earlier, the Church could have suppressed Copernicus's book with ease if it had wanted to. But the printing press made it more difficult to suppress knowledge. Or heresy.

Meanwhile, Martin Luther's challenge to the Church in 1518

The Ptolemaic system required the inventon of epicycles to explain why planets sometimes seem to go backward from their normal orbital paths, as seen from Earth. While Copernicus's heliocentric system made the Solar System seem simpler, it was not until Kepler proved that the planets' orbits are elliptical, rather than circular, that epicycles could be totally discarded.

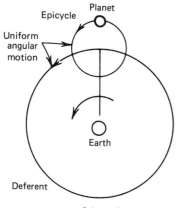

Schematic

had started the Protestant Reformation that would plunge much of Europe into a century of bloody religious war. Rome wanted no part of another challenge to its authority. Still, because Copernicus presented his ideas as an intellectual puzzle rather than palpable fact, the Church did not ban *De Revolutionibus*. At first.

As far as most scholars were concerned, Copernicus's ideas were clearly wrong. The man claimed that the Earth moves! Obviously that is not so. Common sense can tell you that. More than that, though, Copernicus's system did not allow astronomers to

In the Copernican system, the Sun is at the center and the Earth is merely one of its planets. Note that Jupiter is accompanied by four companions, the satellites discovered by Galileo in the earliest telescopic observations of the sky.

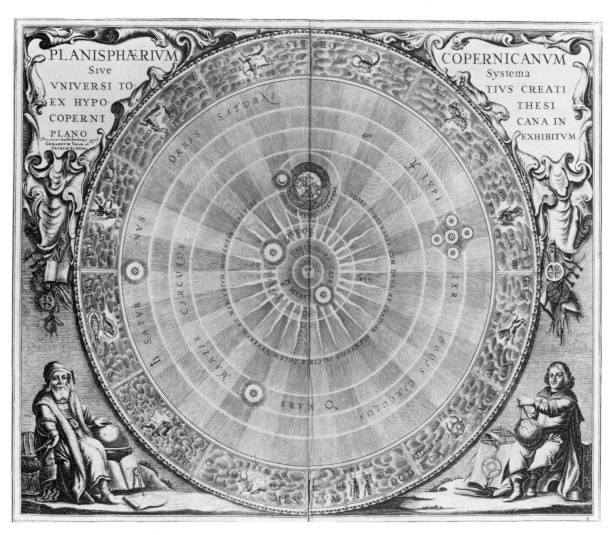

make more accurate predictions of the planets' positions than did the old Ptolemaic system. In the practical world of observations and predictions, there was no clear-cut advantage to Copernicus's newfangled ideas.

The reason is an interesting example of the dangers of false assumptions.

From earliest times, it was assumed that the heavens were perfect and immutable. This attitude prevailed in astronomy for countless centuries, coloring every astronomer's outlook. Because they were part of a perfect and immutable heaven, the stars and planets must move in the most perfect way possible: in circular paths. The circle is the most perfect of all geometrical shapes, it was believed, because every point of its perimeter is exactly the same distance from the center as every other point.

No one bothered to check to see if the planets really do move in circular orbits. Even Copernicus assumed that they did, and as a result, predictions of planetary positions based on his heliocentric theory gave no better accuracies than predictions based on the older geocentric theory.

It took the painstaking work of a Dane without a nose and a German with poor eyesight to finally discover that the planets actually travel in orbits that are almost circular—but not quite.

The Dane was Tycho Brahe, born in 1546, the last and greatest of the naked-eye astronomers. A nobleman who had lost his nose in a youthful duel, the result of an argument over a point of mathematics, he wore a metal prosthesis from the age of 19. Tycho's 40 years of extremely precise observations provided accurate data on planetary motions. What was needed was someone to analyze that data. (See Plate 23 following page 238.)

It fell to Tycho's assistant, the German astronomer Johannes Kepler, to produce that analysis.

Kepler had poor eyesight, which prevented him from becoming much of an observational astronomer. So he turned to mathematics and calculations. He became Tycho's assistant, and when the Dane died in Prague in 1601, Kepler inherited a priceless legacy: the records of Brahe's exquisitely precise observations. Using these data, Kepler worked and sweated and burned the midnight oil—undoubtedly damaging his vision even further—until

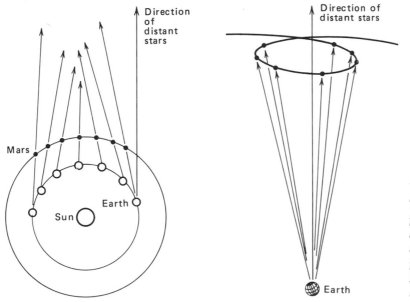

The heliocentric model shows why a planet such as Mars sometimes appears to run backward. It is caused by the changes in relative positions as the Earth "runs past" Mars in its smaller, swifter orbit around the Sun.

he proved conclusively that the planets swing around the Sun in orbits that are elliptical.

And while Kepler was uncovering the basic nature of planetary motion, a very different man ushered in an entirely new era of astronomical study.

In the northern Italian city of Padua, which was then part of the Venetian Empire, Galileo Galilei turned his brand-new telescope to the heavens. As we saw in Chapter 9, the telescope revealed startling new sights. The Moon was not smooth but rugged with mountains and pockmarked with craters. The planet Venus showed phases, like the Moon—which could only happen if Venus revolved around the Sun, rather than around the Earth. The Sun itself was not perfect, as Ptolemaic theory said it must be, but bore spots that came and went and even moved across its flaming surface.

Most staggering of all, Galileo saw the four brightest moons of Jupiter, bodies that clearly revolved not around the Earth but around the oblate disc of that giant planet.

Tycho's observations, Kepler's mathematics, and Galileo's telescope shattered the old Ptolemaic system forever. But not before the Church forced Galileo into silence.

Galileo was a cantankerous old man, and there are those who say to this day that if he had just kept his big mouth shut he could have avoided trouble with the Church. But he could not keep silent. (Why should any of us keep silent, merely because the state or the Church demands it?) Galileo's books brought down the wrath of the Church, and he spent his final years under house arrest.

Like Omar Kayyam six centuries earlier, Galileo held views that ran counter to the teachings of the reigning religious leaders. It

Galileo's early telescopic observations showed that the Sun is not perfect, as it had been thought to be, but was afflicted with spots that moved across its surface.

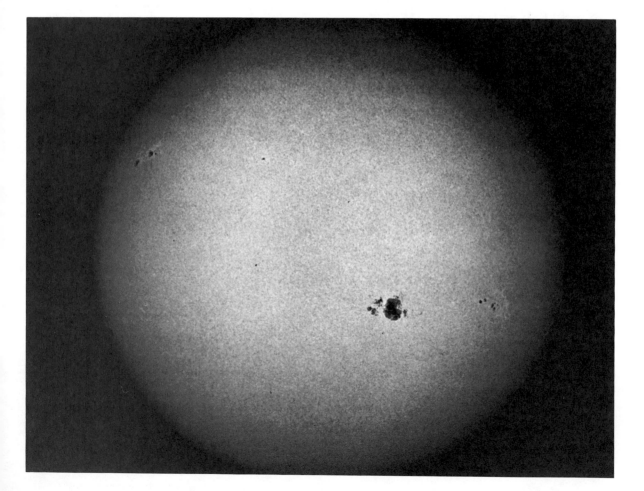

seems inevitable that when political power falls into the hands of a church, the most conservative members of that church eventually gain control and try to suppress all views that are not their own. It has happened more than once to Islam and to Christianity as well. When Christianity became the official religion of the Roman Empire, for example, the newly powerful Church closed the philosophical schools of Athens, the intellectual descendants of Socrates and Plato. The decline of Roman thought, and power, soon followed.

The basic difference between science and religion is the difference between faith and skepticism. The religious zealot *knows* the truth and will brook no countering opinions. As George Bernard Shaw put it, a fanatic who is willing to die for his cause will think nothing of killing you for his cause. On the other hand, the scientist seeks knowledge and is always ready to consider new evidence that may junk previous ideas. The most dearly held convictions, such as Newton's belief in the particle nature of light, can be overthrown by fresh evidence. Science is a process, and scientists are constantly changing their opinions in light of new observations. Religious fanatics cannot abide such uncertainty, such constant questioning, such refusal to accept an immutable "truth."

The Inquisition threatened the 70-year-old Galileo with torture if he did not publicly recant his heretical view that the Earth revolved around the Sun. Broken in health and spirit, going blind, Galileo bowed to its overwhelming force.

Legend has it, though, that as he left the chamber in which he made his recantation he muttered, *"Eppur si muove."* "And yet it does move."

Authority and force could not stem the tide of new knowledge. The lights in the sky that had been thought to be the abode of the gods were becoming something much more exciting. They were becoming a vast classroom in which astronomers and physicists could learn how the universe works, what it is made of, and—most thrilling of all—how it began.

20

Fingerprints from Rainbows

THE YEAR that Galileo died, 1642, Isaac Newton was born. He was a premature baby and barely survived his first few weeks. What would the world have done without him?

You recall from Chapter 9 that Newton's concept of universal gravitation was the first realization that forces at work here on Earth also work in the heavens, indeed throughout the universe. The force of gravity that makes an apple fall to the ground also keeps the Moon in its orbit and—as astronomers later found—holds together whole galaxies of billions of stars.

Important as his theory of gravitation is, Newton's work on optics led to even greater contributions to our understanding of the universe. He invented the reflecting telescope, a major tool for observing the heavens. More than that, however, Newton's simple experiments with light and prisms led to the most powerful instrument astronomers have for prying into the secrets of the stars.

Scientists are hunters, detectives who track down the elusive workings of nature. If astronomers can be pictured as detectives, then their suspects and witnesses are the points of light in the sky that we call stars. How do you interrogate a suspect that is trillions of miles—light-years—away from you? All you can see of even the nearest stars is a pinpoint of light. (Except, of course, for the Sun.)

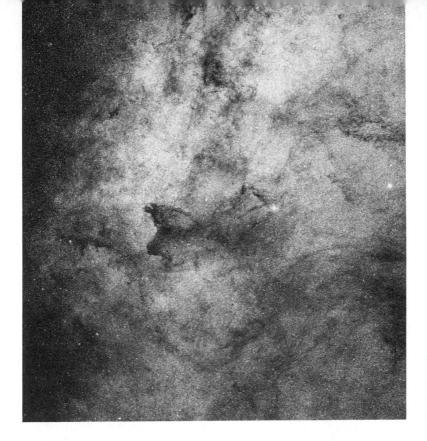

Although telescopes have allowed us to see stars beyond counting, for the most part each star remains nothing more than a pinpoint of light in the astronomer's eyepiece.

Even in powerful telescopes the stars remain dimensionless points of light. This is often a crushing disappointment to the new amateur astronomer with a backyard telescope. I have an eight-inch reflector, and although friends are awed and delighted by viewing the craters of the Moon or the Galilean satellites of Jupiter, they are invariably surprised when they realize that the stars appear no bigger than they do to the unaided eye. Telescopes allow us to see more stars, to peer deeper into the vast darkness. But each star remains a tiny pinpoint of light.

That is where Newton and his prism come in. The rainbow pattern of the spectrum, it turns out, is a kind of fingerprint that can tell an astronomer-detective a truly astounding amount of detailed information about a star or a cloud of gas shining in deep space.

Light from a star, or from any natural or artificial source, can be spread into a spectrum of separate colors by using a prism or a device called a diffraction grating. Astronomers use such devices in *spectroscopes* attached to their telescopes. The spectrum is invariably recorded on photographic film for analysis.

Newton discovered that "white" sunlight contains all the visible colors of the rainbow. Later scientists found that the Sun and stars emit a great deal of electromagnetic energy that is invisible to our eyes: infrared, radio waves, ultraviolet, X-ray, and gamma-ray energy. *Any* physical body that is above absolute zero in temperature (in other words, everything in the universe) radiates electromagnetic energy in one form or another. Your own body, running at approximately 98.6° Fahrenheit, is radiating in the infrared frequencies. IR-sensitive "snooper scopes" can see this infrared glow even in absolute darkness.

In Germany in 1859 the physicist Gustav Kirchhoff and the chemist Robert Bunsen* uncovered the basic laws that govern the way bodies radiate light.

Kirchhoff and Bunsen found that when a dense object is heated to the point at which it will glow, it gives off a rainbowlike blend of many colors. This is called a *continuous spectrum*. The dense object could be the red-hot tip of a soldering iron or the massive body of the Sun or a star.

Sunlight, the white light radiating from the dense body of the Sun, is essentially a continuous spectrum, a blend of colors from deep red to far violet that we see in the rainbow when raindrops in the air create a natural prism.

When white light is passed through a thin gas, something different happens. While the white light itself shows a continuous spectrum, when it goes through the gas, dark lines appear in its spectrum. It is as if something in the gas is stealing thin slices away from the continuous spectrum. In fact, that is exactly what is going on. The atoms or molecules of the gas are absorbing certain selected wavelengths of the white light, leaving dark slits in the spectrum at those particular wavelengths. This is called an *absorption spectrum*.

Different elements or compounds in the gas absorb their own particular wavelengths of the white light at specific locations in the spectrum. Each chemical element or compound produces its own characteristic set of absorption lines, as unique as a fingerprint. This makes it possible to identify the elements in a gas, whether

---

*The same Bunsen who invented the Bunsen burner, a standard heating tool for chemistry laboratories.

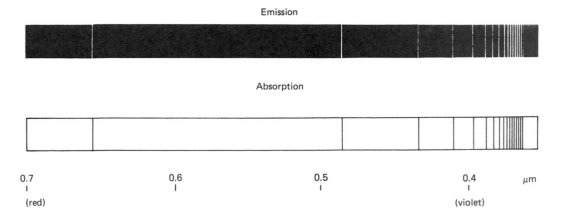

Absorption

0.7   0.6   0.5   0.4   μm

(red)   (violet)

it is in a laboratory, or pouring out of a smokestack, or shining out in deepest space.

The surface of the Sun is called the photosphere. It is a region of fairly thin gas lying atop the dense, bright light source of the deeper layers of the glowing star. That deeper, dense main body of the Sun emits a continuous spectrum, a broad rainbow of every visible color. But as that white light comes up through the thinner layer of the photosphere, some 600 absorption lines are scratched across it, so that when astronomers look at the Sun's spectrum, they see an absorption spectrum. The German physicist Joseph Fraunhofer made one of the earliest studies of these lines in 1814, and to this day they are still called Fraunhofer lines.

What do the Fraunhofer lines tell us? Scientists went to their laboratories and shone white light through every type of gas they could create. They built up a catalog of the wavelengths absorbed by all the known elements. By comparing the wavelengths of absorption in their laboratory specimens with the wavelengths of the Fraunhofer lines, they soon identified some 60 elements in the Sun's photosphere. For the first time, scientists could tell the chemical composition of a star. Part of it, at least.

A single element will produce more than one absorption line, so the 60 elements identified in the Sun's spectrum accounted for almost all of the 600 lines. But there were some surprises.

The solar spectrum revealed a new element, never before seen on Earth. It was called helium, in honor of the Sun. Identified in 1868, helium was discovered on Earth 27 years later in the

Emission spectra (top) show bright lines in an otherwise dark spectrum. Absorption spectra show dark absorption lines against a continuous spectrum of rainbow-like light. Each line is characteristic of a chemical element, and both emission and absorption spectra are used to identify elements in stars, interstellar clouds, and galaxies.

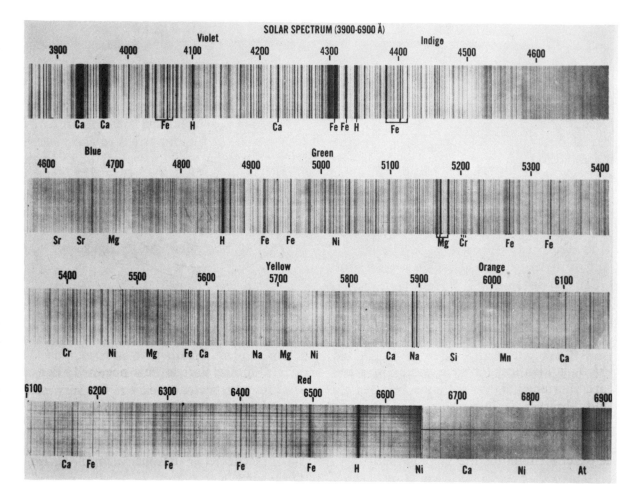

SOLAR SPECTRUM (3900-6900 Å)

The Sun's spectrum, showing the colors of the light, the wavelengths in Ångstrom units, and the atomic elements that create the dark Fraunhofer lines of absorption.

by-products of the radioactive breakdown of uranium. It was the first—and to date the only—element discovered in the stars before being found on Earth.

Kirchhoff and Bunsen also found that under certain conditions a gas will emit bright lines of color, rather than a broad, continuous spectrum. This is called an *emission spectrum*, and here too the spectral lines can be used as fingerprints, since each element emits only at specific wavelengths. Bright clouds of gas shining in deep interstellar space show emission spectra, and astronomers can determine their chemical composition with the spectroscope.

Using stellar spectra, astronomers have been able to determine the chemical compositions, temperatures, sizes, brightness, motion, and even the rotation rates of stars, interstellar gas clouds, and entire galaxies containing hundreds of billions of stars.

Before we can see how they make these determinations, we must briefly look at how astronomers measure the distances to the stars. Distance measurements are crucial—and the weakest link in astronomical knowledge.

George Washington, in his days as a surveyor, knew how to measure the distance to an object he could not physically reach. To measure the width of an uncrossable river, for example, he would stretch out his surveyor's chain and, using it as a baseline, measure the angles between each end of the chain and an object on the other side of the river—say, a tree on the opposite bank. The tree served as the apex of a triangle, the chain as the triangle's base. Knowing the length of the chain and the two base angles, it was a simple matter of trigonometry to determine the distance from the chain to the tree.

This trigonometric method is used by astronomers to measure the distances to the stars. The baseline, in this case, is the diameter of the Earth's orbit around the Sun, a baseline that is 186 million miles long. Barely long enough, because the stars are fantastically far away.

The astronomer sights on the star he or she wants to measure and notes the position of that star against the fainter background stars that are presumably much farther away. (If they are not farther away, the measurement is ruined.) Six months later the astronomer squints at the star again and measures its shift of position against the background stars. This is called the *parallactic shift*. It is similar to the shift you see when you extend your arm, hold up your thumb, and then view the thumb against the background landscape first with one eye shut and then the other. The thumb seems to hop back and forth against the background as you switch eyes.

The astronomer fervently hopes the star he or she is measuring will move back and forth against the fainter background stars. That parallactic shift gives the astronomer the angle at the apex of a *very* long, thin triangle. Knowing the apex angle and the length of the baseline, the distance to the star can be calculated.

Angles are measured in degrees, minutes, and seconds. One

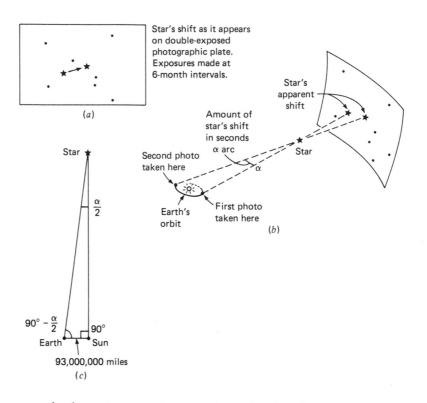

(a)

Star's shift as it appears on double-exposed photographic plate. Exposures made at 6-month intervals.

(b)

(c)

Measuring the distance of a star by trigonometric parallax: (a) shows the apparent six-month shift of the star's position as seen against the unmoving background of farther, fainter stars. (b) shows how Earth's motion around the Sun causes the apparent shift. The amount of shift seen determines the size of the angle. (c) shows how a triangle can be constructed from the Earth to Sun to star. With the length of the Earth-Sun baseline known and all three angles known, the distance to the star can be calculated.

second of arc is an angle of 1/3,600th of a degree, the angle that a 25-cent piece would show—at a distance of three miles. If a star showed a parallactic shift of one second of arc, its distance would be 3.26 light-years.

There is no star close enough to us to show a parallactic shift of one second of arc. All the parallaxes that astronomers measure are smaller than that. The nearest star is the triple system of Alpha Centauri, which consists of two stars rather like the Sun orbiting closely around each other and a third component, a red dwarf star that swings around the other two in a wider orbit. Since the red dwarf is slightly closer to our Solar System than its two bigger companions, it is often called *Proxima* Centauri. The Alpha Centauri system shows a parallax of about 0.76 second of arc, which makes its distance 4.3 light-years.

The angles measured by astronomers are exceedingly small, and thus they have been able to measure the distances to only a few hundred of the closest stars using the trigonometric method. For more distant stars they can only estimate distances by comparing brightnesses.*

Incidentally, astronomers use the term *parsec* as a unit of distance for two reasons. We have already seen that the light-year is a distance unit equivalent to the distance that light travels in one year: roughly 6 trillion miles. But people often confuse light-years with a measurement of time, and the light-year is a bit too small for truly long-distance measurements. So astronomers use the parsec, the distance given by a *par*allactic shift of one *sec*ond of arc: 3.26 light-years. Think of the light-year as the astronomer's equivalent to a foot, the parsec as the astronomer's yard.

Now we can see how much astronomers have been able to learn from the pinpoint lights of the stars by combining spectroscopy with distance measurements.

Chemical composition of the stars and interstellar gas clouds is learned by examining the absorption or emission lines of the spectra. Each line is produced by a specific chemical element or compound. It turns out that the observable universe is almost entirely hydrogen and helium, the two lightest elements, with only about 1 percent of heavier elements. Our planet Earth, you, and I are composed mainly of "impurities" in the hydrogen-helium makeup of the cosmos.

Temperature is determined through a relationship discovered in 1893 by the German physicist Wilhelm Wien. You can roughly gauge the temperature of a flame by looking at its color. A blue flame is hotter than a yellow one, which in turn is hotter than a red flame. The farther toward the blue end of the spectrum, which is the more energetic end, the higher the temperature of the flame.

Wien found a more precise mathematical relationship between temperature and color. The wavelength at which most of the star's energy is being emitted is proportional to the temperature of the star. Determine the most intense wavelength of the star's spectrum and you can calculate its surface temperature.

Once you know a star's distance you can determine how bright

*Actually there are certain variable stars in the sky whose distances can be estimated by timing the changes of their brightness. While this technique has allowed astronomers to estimate very large distances, it is far from foolproof.

it is by comparing its apparent brightness against its distance. Brightness decreases with the square of distance: A light that is moved twice as far away as it originally was will appear four times dimmer; move it three times as far and it will appear nine times dimmer. By determining the star's distance, it is possible to tell how luminous the star truly is.

Once you know a star's surface temperature you can calculate how much energy it is radiating per square centimeter of surface area, then compare that figure to its true luminosity. This will tell how many square centimeters of surface area the star has; in other words, its size.

If a star is moving across one's field of view its motion can be measured simply by being patient. Tycho found that a few stars were not where Ptolemy's *Almagest* had said they should be. In the 14 centuries between the two astronomers' lives, those few stars had moved far enough to be observably displaced. All the other 2,000 or so naked-eye stars had moved too, but they are so far away that their motions made no observable difference from Earth. Astronomers call this kind of sideways motion across the sky *proper motion.*

Spectroscopic measurements can determine a star's motion along the line of sight whether the star is approaching or receding from the observer. This is thanks to the *Doppler effect.* Christian Doppler, a nineteenth-century Austrian physicist, explained why a sound increases in pitch as it approaches the listener and then decreases in pitch as it moves away. You have heard this whenever a fire engine or ambulance wails past. In 1848 Doppler suggested that light would behave the same way, and French physicist Armand Fizeau proved him to be right.

A source of light approaching the observer will show a shift in its spectrum toward the blue end. A light source that is receding will be shifted toward the red end of the spectrum. This holds true for *any* source of light: a candle, a star, or a galaxy. Slight shifts in a star's spectrum can tell astronomers whether the star is moving toward or away from us.

As they studied stellar spectra, astronomers found that the absorption or emission lines of some stars moved back and forth over time. They realized that this movement was caused by the star's rotation, and by plotting the shifts of the spectral lines they

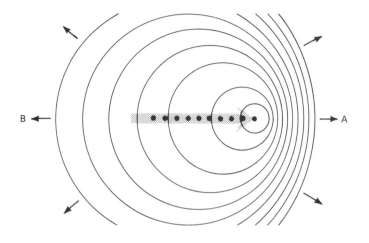

The Doppler effect. The gray arrow shows the direction of movement of a wave source. Each dot represents the center of a circular wave emitted by the moving source. As the source moves, waves "pile up" at A, spread out at B. The waves could be light or sound; if light, A represents the blue end of the spectrum, B the red.

could determine the star's rate of spin. The Sun turns slowly, about once every four weeks. It does not rotate as a solid body; the equatorial regions turn more slowly than the polar regions. This is proof that the Sun is not a solid body but gaseous.

Other stars rotate much more rapidly, and *pulsars,* which are stars that have collapsed down to a diameter of only a few miles, spin around in a mere second or so!

Spectral analysis is so important in astrophysics that astronomers have developed a system for classifying the stars by their spectra. The system is based on stellar surface temperatures. Originally the system was in strict alphabetical order—the hottest stars were labeled class A, the next hottest B, and so on. But continuing refinements to the observations showed that some of the classifications were out of line. B stars turned out to be hotter than A stars, for example, and a new class of stars was discovered that was hotter still. The classifications were shuffled back and forth, some classes were dropped altogether, and all semblance of alphabetical order was lost.

The present classification scheme, starting with the hottest known stars and going to the coolest, reads: O, B, A, F, G, K, and M, plus four classes of rare types of star, W, R, N, and S.

Astronomers have no trouble remembering the order. They simply mutter the mnemonic "Oh, be a fine girl, kiss me!" Some add, "Right now—smack!" Apparently they remember the W without any help.

STARS

Spectral Class	Surface Temperature (K°)	Color	Examples
O	above 25,000	blue-violet	rare
B	11,000–25,000	blue	Rigel, Spica
A	7,500–11,000	blue-white	Sirius, Vega
F	6,000–7,500	white	Canopus, Procyon
G	5,000–6,000	yellow	Sun, Capella
K	3,500–5,000	orange	Arcturus, Aldebaran
M	below 3,500	red	Betelgeuse, Antares

Temperatures given in degrees Kelvin, where 0 equals absolute zero and the freezing point of water is 273°.

The Sun is a G-class star, yellowish in color, quite stable and long-lived. By way of comparison, listed in the table are the spectral classes, surface temperatures, and colors of some of the brighter stars that you can see on almost any clear night.

Galileo's original little 30-power telescope has evolved into giant observatories where immense telescopes peer out toward the edge of the observable universe. The mecca of optical astronomers is Mount Palomar in southern California; there stands the 200-inch reflector telescope named after the astronomer George Ellery Hale. It is not the largest telescope in the world; there is a 236-inch, though inferior, instrument in Russia. But the Mount Palomar observatory is truly a monument to human intelligence and curiosity. Astronomers from all over the world come to it to seek the answers to the grandest and most puzzling questions of the universe.

A decade or so ago most astronomers were convinced that the Mount Palomar telescope (and its slightly larger cousin in the USSR) represented the last of the giant optical instruments to be built on Earth. There are limits to how large an optical telescope can get, limits set in part by the turbulence of the Earth's atmosphere.

The air that looks so clear to us is a murky muffling blanket to the optical astronomer. Our atmosphere is hazy and constantly in motion. That is why astronomical observatories are usually perched on mountain tops: to get above as much of the clouds and weather as possible. Still, the atmosphere's fundamental turbulence places limits on the useful size of optical telescopes. Even atop the tallest mountain the air is not still. This constant turbulence is what causes the stars to twinkle. To an astronomer with a very sensitive telescope this twinkling effect is magnified to the point where the star being observed fades in and out of focus and even jumps entirely out of the telescope's field of view. And the bigger the telescope, the more sensitive it is to such problems.

But astronomers are an ingenious lot, and they have recently come up with several new ideas for ground-based optical telescopes.

The first of these was the Multiple Mirror Telescope (MMT) developed by the University of Arizona and the Smithsonian Astrophysical Observatory. Instead of using one large mirror, the MMT employs six mirrors, each about 72 inches wide. They are linked through a computer to focus properly on a single pinpoint of starlight. The MMT is equal in light-gathering power to a reflector of more than 170 inches' diameter.

J. Roger Angel, of the University of Arizona, has been at the forefront of developing even newer technologies for optical telescopes and foresees reflectors with mirrors of more than 300 inches.

The European Southern Observatory (ESO) organization is planning its Very Large Telescope facility, which will consist of four eight-meter (315-inch) telescopes standing side by side along a distance of 104 meters (about 340 feet). The $240 million facility will be built in Chile's high Atacama desert. ESO is a consortium of eight nations: Belgium, Denmark, France, West Germany, Italy, the Netherlands, Sweden, and Switzerland.

In the meantime, two new technologies have entered the field of optical astronomy, creating the greatest changes since 1609: the technologies of electronics and space flight.

Electronic boosters have been added to telescopes. Based on Einstein's work on the photoelectric effect, the first electronic boosters used photoelectric cells, similar to the electric eye that opens supermarket doors when you step through a beam of light. More recently, *charge-coupled devices* (CCDs), solid-state chips somewhat akin to semiconductor diodes, have been applied both to astronomical telescopes and to the microscopes used by biologists and medical researchers.

Basically, the photoelectric cell and the CCD convert light energy into an electrical current. The simplest use of the photoelectric cell is as a photometer, or light counter: X amount of light produces X amount of electrical current; twice X of light yields twice X of electricity, and so on. There's a photometer in your automatic camera, a tiny chip that serves as a light meter and warns you when there is not enough light to obtain a good photograph. Astronomers use photometers to make precise measurements of the amount of light being received from a star.

To measure the light from very faint sources, astronomers use photomultipliers. Basically the same as a photometer, the photomultiplier is designed to produce a predictably multiplied electrical current for a given input of light. X amount of light can yield 10X amount of current, or 1,000X, or even a million times X. With the photomultiplier's power of "magnification" known and under careful control, very faint objects can be studied and their light output determined precisely.

Photometers and photomultipliers give their information in the form of a meter reading, a pointer on a dial, a number. It was only natural, to a generation raised on television, to adapt such electronic technology to produce images, pictures, and optical data that can be played on a TV screen and recorded on videotape.

In the old days (meaning, when I was younger) astronomers shivered their nights away in the unheated domes of their observatories. Heating the dome would have meant bubbling currents of warm air rising past the telescope, ruining its ability to see clearly. And of course the best nights for observing the stars are those cold still nights when a blanket of arctic air lies across the land.

Today, astronomers sit in cozy warm offices and watch a TV screen. Well, some of them do.

The light coming into the telescope is boosted (astronomers usually say "enhanced") by electronic devices not unlike the basic photomultiplier. The electrical current is then fed into a video system to produce an optical image. With electronic enhancement a telescope can effectively increase its light-gathering power many times. A 40-inch telescope can be boosted to the sensitivity of a 100-inch instrument.

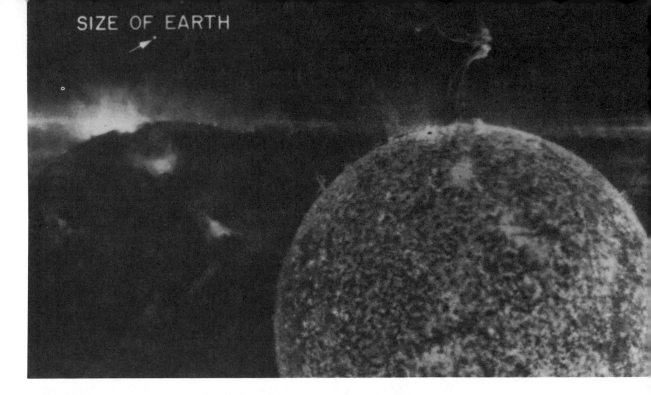

SIZE OF EARTH

Meanwhile, the advent of space flight has allowed astronomers to get above the atmosphere entirely. For optical astronomy, this is as big a step forward as Galileo's telescope was over naked-eye observations. But there is more: The atmosphere blocks out many wavelengths of electromagnetic energy. Once in orbit beyond the atmosphere's screening effect, astronomers can study the *entire* electromagnetic spectrum, from radio waves through infrared, ultraviolet, X-rays, and gamma rays. (See Plate 24 following page 238.)

Whole new fields of astronomy have opened up thanks to space flight. High-energy astronomy studies the UV, X, and gamma radiation coming from the stars. A single satellite called IRAS, for Infrared Astronomy Satellite, produced enough data on the IR wavelengths to keep hundreds of astronomers busy for decades to come. IRAS pictures even showed a few stars that are surrounded by clouds of dust and other particles—the beginnings, perhaps, of planets.

Large optical telescopes are going into space. The largest of them, the Hubble Space Telescope, is a 94-inch instrument that may be able to detect actual planets orbiting other stars. Astrophysical theory leads to the conclusion that many stars have planetary systems revolving around them. But the stars are so distant, and planets are so small and dim, that no such planets can be

Portrait of a violent eruption on the Sun, taken at ultraviolet wavelengths by astronauts aboard the Skylab space station in 1973. The white dot represents the size of Earth on the same scale.

Radio telescopes have opened up an entire new field of astronomy since 1945. Modern "dishes" are even used to search for possible signals from extraterrestrial intelligences.

seen from Earth. The Space Telescope has the power to pick out planets that may be in orbit around the nearest stars. If they are there.

While our atmosphere blocks out almost all the wavelengths of electromagnetic energy except visible light, and a little infrared and ultraviolet on each end of the visible spectrum, it also allows some radio wavelengths to penetrate to the ground.

Radio astronomy started in 1932, when a young engineer working for the Bell Telephone Laboratories, Karl Jansky, was assigned to track down the source of electrical interference ("static") that often bedeviled long-distance radio communications. Jansky soon found that the source came from the stars.

World War II put an end to almost all astronomical research, but it also forced rapid growth in the fledgling technology of electronics. After the war, radio astronomy blossomed into a full partnership with optical astronomy, and radio astronomers made some of the most startling discoveries of all.

The quasars, the pulsars, and the faint echo of the universe's original Big Bang were all first detected by radio telescopes.

21

The Starry Messengers

WE CAN NOW embark on a brief tour of the universe. Our tour guides will be the light beams given off by the stars themselves. We will examine the messages that the stars are sending us, messages that will take us to the very edge of the universe and the beginning of time.

Please fasten your seat belts.

Astronomers began with their unaided eyes. In 1609 they started using telescopes. To these instruments, over the years, they added cameras, spectroscopes, and electro-optical boosters. In the 1940s they began to observe the universe with radio telescopes. By the 1960s they were opening new doors with instruments in space capable of studying all the electromagnetic wavelengths that are blocked by our atmosphere.

Astronomy has spawned astrophysics. The astronomer observes the universe; the astrophysicist tries to determine what makes it work the way it does. As you might suspect, the division between the two allied disciplines is not strict. Many astronomers are also astrophysicists, although many physicists who have dealt with astrophysical problems have never been observational astronomers.

It was one of those nonastronomer physicists who discovered what makes the Sun shine. Actually, an astronomer made the same discovery at about the same time, but he has not shared in the glory as much as the physicist.

By the 1930s, astronomers had determined that the Sun was a huge glowing sphere, slightly more than 866,000 miles wide (nearly 110 times the diameter of the Earth), with a surface temperature of 5750°K. Thanks to spectroscopic analysis of the Sun's light, they knew that the Sun was composed almost entirely of hydrogen and helium, with only a few percent of heavier elements such as carbon, oxygen, and so on.

They believed that the Sun had been shining steadily for billions of years, although they had no positive way to verify this. Geological studies of the Earth showed conclusively that our planet is more than 4 billion years old. Paleontologists can trace the origins of life 3 billion years into the past; the Sun must have been shining steadily during that time, or life would have been wiped out. Meteorites, chunks of stone and metal that hit the Earth from space, also gave evidence of being some 4 billion years old.

When astronomers began to realize, in the nineteenth century, how old the Sun must be, they were faced with a crucial question: If the Sun has been shining for more than 4 billion years, what is the source of its incredible energy? They could measure how much energy the Sun radiated. In modern terms, it comes to the equivalent of 10 billion megatons of TNT exploded every second. Ten billion megatons per second! For billions of years!

How much energy is that, in terms we can understand from everyday experience? The astronomer George Abell put it this way:

> [Picture] a bridge of ice 2 miles wide and 1 mile thick, and extending over the nearly 100-million-mile span from the earth to the sun. If all the sun's radiation [energy] could be directed along that bridge, it would be enough to melt the entire column of ice in one second.

No one could account for this prodigious energy. If the Sun were a mixture of pure carbon and oxygen and burned like a coal flame, it would have burned itself into a giant ash within a few thousand years. Some astronomers suggested that the Sun was slowly shrinking and converting gravitational energy into heat and light. But calculations showed that such an energy source would be played out in only a few million years.

In 1905, Einstein (there's that name again) published his fa-

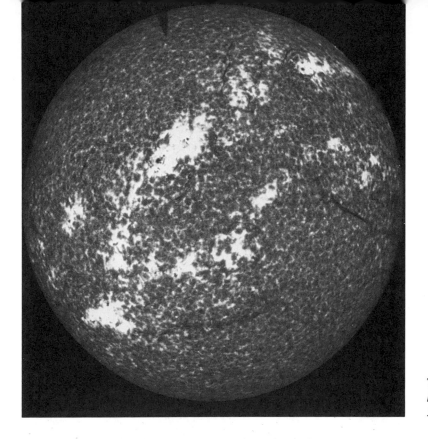

The Sun, as photographed in the light of calcium's absorption line at 3,934 Ångstroms.

mous $E = mc^2$ equation. Astrophysicists quickly realized that the Sun must be a titanic nuclear reactor, converting the immense energies in atomic nuclei into light. Yes, but how?

Hans Bethe, a German physicist who had fled to America to escape Hitler, produced the answer in 1938. At the same time, Carl von Weizsacker, an astronomer who remained in Germany, worked out the same answer. The two men were separated by thousands of miles and worked quite independently of each other. It often happens in science that two researchers will hit upon the same solution to a problem simultaneously and independently, usually because both are working on the same problem and following along the same path. As Mark Twain put it, "When it's steamboat time, you steam."

However, Bethe was awarded the Nobel Prize in 1967 for his work on determining the origin of the Sun's energy. Von Weizsacker was not.

The astrophysicists determined the principles of the hydrogen fusion process, which we first looked at in Chapter 16. Although no one could peer into the interior of the Sun to directly examine what is going on there, astrophysicists were able to calculate what the conditions at the Sun's core must be. They had determined the Sun's mass to be some 2×10^{27} tons (2 billion billion billion tons). The weight of that enormous amount of matter produces pressures at the Sun's core that are a billion times the pressure of the air we breathe and temperatures of 20 million degrees.

Those are the conditions at the heart of a modest-sized star of medium temperature. In that raging inferno, atomic nuclei are forced to fuse together. Essentially, four hydrogen nuclei fuse into one helium nucleus, yielding energy in the process. Bethe worked out the complex reactions that are now known as the *carbon chain,* in which carbon nuclei act as catalysts to promote the hydrogen fusion.

In stars larger and hotter than the Sun, the direct *proton-proton* reaction prevails, where hydrogen nuclei (simple protons) interact directly without the need of a catalyst.

As we saw in Chapter 16, the hydrogen fusion process converts 0.7 percent of the original mass of the hydrogen nuclei into energy. The energy that we see as sunlight begins deep in the core of the Sun, where the fusion processes are taking place. Hydrogen fusion produces photons of extremely energetic gamma-ray wavelengths. But the ionized gas that makes up the Sun's interior is so dense that the photons cannot travel very far before they interact with other particles. Each interaction leads to the production of a slightly less energetic photon.

By the time the energy of an individual fusion reaction reaches the Sun's surface, it is no longer a gamma-ray photon; it has been "eroded" into a photon in the visible spectrum, a chunk of "white" light. And a good thing, too! A star that emits gamma rays would fry the surface of any nearby planet.

It takes something like a million years for a photon to worm its tortuous way from the Sun's core to its surface. From there, it takes only eight minutes to travel through the 93 million miles of interplanetary space and reach the Earth. So the sunlight we enjoy today began as gamma-ray energy at the heart of the Sun about a million years ago.

The Sun converts some 4 million tons of its matter into energy every second and has been doing so for nearly 5 billion years. It will be able to continue doing this for another 5 to 10 billion years. If the Sun should ever consume all its hydrogen, it will be only 0.7 percent lighter than it is now.

But the Sun will probably never reach that point of consuming all its hydrogen. Long before that time arises, the Sun will begin to change so drastically that life on Earth will become impossible.

We know about the stages that the Sun will go through because astronomers can see a plethora of stars in the heavens at all stages of life. Stars have life cycles, just as people do, although a star's life is measured in billions of years. Stars are born, live out a certain span of eons in peaceful glowing equilibrium, then enter a turbulent phase that finally ends in—well, the possible ending points of a star's life are so strange that they cannot be easily summarized. Have patience—we shall see them soon enough.

There are so many stars in the heavens that astronomers can see every stage of stellar lifetimes from birth to death. They see stars that pulsate, stars that collapse, stars that explode. Some stars swell to such enormous size that they would swallow up any planets orbiting around them. Others are being born out of swirling clouds of interstellar gas and dust.

There are stars of all colors, giant stars, dwarf stars, double and triple stars. Our Sun is a moderate, middle-of-the-road star; yet it too will someday begin to swell and then collapse.

In 1948 a trio of British astrophysicists uncovered the fundamental processes that govern the stars' life cycles. Hermann Bondi, Thomas Gold, and Fred Hoyle announced their Steady State Theory that year, a cosmological theory that proposed a novel answer to the biggest question of them all: the origin of the universe. As we shall soon see, most astronomers did not accept the Steady State view of how the universe began. But built into the theory was a description of how stars work. That has been tested and proved.

Stars are born out of clouds of interstellar gas and dust. Von Weizsacker and others showed how swirling eddies in such clouds can produce a star and—perhaps—planets orbiting around it. Astronomers have photographed *protostars*, dark clumps of gas and dust. (See Plate 25 following page 238.) Some protostars, photographed repeatedly over a span of years, have been caught in the very act of beginning to shine. Astronomers have seen the actual birth of stars.

Picture a protostar, a clump of swirling gas and dust about a light-year in diameter. The gas is composed of about 90 percent hydrogen, a few percent helium, and some traces of heavier elements. The cloud is dark and cold; astronomers can only see those that are backlit by glowing clouds of interstellar gas. As the protostar rotates, its matter sinks in toward the center. Density and temperature at the center begin to rise. All those megatons of matter pressing inward bring the central temperature up to many millions of degrees. The protostar is glowing a dull red now, the heat created by gravitational contraction.

At some point, the temperature and density at the core of the star become so high that the hydrogen nuclei begin to fuse together to create helium and energy. The star "turns on," and its light is now truly starlight.

For billions of years the star can glow steadily. It has achieved a fine equilibrium. Gravity still wants to pull all those tons of matter in toward the center, but the outward-pushing forces of gas and radiation pressure balance gravity and keep the star at a steady size.

This can go on as long as the hydrogen fusion process keeps producing energy. This is the stage that the Sun is in now, the mature sedate adulthood of a star. Unlike humans, stars go from birth to maturity rather quickly in terms of their total life span. A protostar may turn on and settle into maturity in a scant hundred thousand years or so.

It is only after a long steady adulthood that a star becomes turbulent. In human terms, the teenage years come *after* adulthood.

Although stellar life spans are measured in billions of years, sooner or later a star will deplete its hydrogen fuel. That is when its turbulent years begin. Astrophysicists have found that as the hydrogen is consumed, the star's core becomes mostly helium—the by-product "ash" of hydrogen fusion. Helium is denser than hydrogen, and as the core turns into helium its density increases. This makes the core hotter. Contrary to what our earthbound common sense would suggest, the star's core becomes hotter as its hydrogen fuel is consumed.

When the helium content of the core becomes high enough, the central density becomes so great that the star's core temperature rises to some 100 million degrees. At 20 million degrees hydrogen fuses into helium. At 100 million degrees helium begins to fuse into heavier nuclei of carbon, oxygen, and neon.

The star has found a new energy source. But with a much hotter central temperature, the outer layers of the star are pushed by gas and radiation pressure harder than they had been before. The star swells. Gravity yields and the star grows. When this time comes to the Sun, our friendly yellow star will bloat into a reddish giant and its outer envelope will swell to engulf the orbits of its nearest planets.

Earth's atmosphere and oceans will boil away. Life will become impossible on our planet. This will happen in 5 to 10 billion years.

But while life ends on Earth, for the Sun this is merely the start of a new phase in its life cycle.

The helium fusion reactions run much faster than hydrogen fusion, and soon (in stellar terms) the helium supply at the star's core runs low. Again, the core has been increasing in density and temperature because now it is composed of elements that are much heavier than the original hydrogen. Like a narcotics addict looking for an ever-more-potent fix, the star begins fusing oxygen, carbon, and neon into still-heavier elements.

The cycle keeps repeating but in ever-tightening coils: Fusion reactions make heavier elements that increase the core's density, which makes the temperature rise, which leads to new fusion reactions that produce still-heavier elements. Each new cycle goes faster than the one preceding it. The star is behaving like the Hemingway character who, when asked how he went bankrupt, replies, "Two ways . . . gradually and then suddenly."

The star is heading for disaster. New energy sources will not be forthcoming forever. Outwardly the star continues to swell. At some point, certain stars begin to pulsate. Gravity is fighting a battle now against the outward-pushing gas and radiation pressures.

Finally the star's core consists largely of iron. Nuclear fusion reactions based on iron produce *lighter* elements, not heavier ones. The star has reached the end of its tether. It cannot produce more energy in its core.

Suddenly the gas and radiation pressures that had been pushing the star's matter outward from the core disappear. Gravity is still there, however, and the star collapses in on itself.

What happens now depends on the star's original mass. For a star of the Sun's mass, the collapse shrinks the star to a hundredth of its original size. The Sun will eventually become a seething ball of plasma about the size of Earth. A slim skin of nuclear reactions may continue to simmer near the star's surface. Such stars have been observed. Astronomers call them white dwarfs. They are high in temperature, small, and so dense that a spoonful of their matter would weigh thousands of tons.

Gradually a white dwarf will cool off and turn dark. Such is the probable fate of the Sun tens of billions of years from now.

The brightest star in our night sky, Sirius, is accompanied by a white dwarf companion. Since Sirius has been known from time immemorial as the Dog Star, astronomers have dubbed its tiny companion the Pup.

More massive stars have more spectacular death throes. When the outer layers of a massive star collapse into its iron-rich core, it triggers the titanic kind of explosion that astronomers call a *supernova.* A supernova can release a billion years worth of stellar energy in one titanic blast. The star flames into a radiance some 100 million times brighter than the Sun. Such an explosion would utterly destroy any planets nearby.

But it does not destroy the star's core. The core remains, collapsed down to a few miles in diameter. The density of the remaining matter becomes so incredibly high that the electrons are literally squeezed into the atomic nuclei to combine with their protons and create a star that consists entirely of neutrons. Such *neutron stars* spin on their axis in times measured in seconds. Some of them emit pulses of radio and light energy.

These *pulsars,* as they are called, caused a flurry of excitement in 1967 when radio astronomers first detected their signals. Nothing like these signals had ever been found before. The pulses came in bursts of a few thousandths of a second and were timed as precisely as the most accurate clocks on Earth. Some astronomers believed they might be deliberate signals sent out by an intelligent civilization, and for weeks the LGM (Little Green Men) theory competed with other ideas about what could be causing the radio pulses.

It turned out that the pulsar signals were coming from rapidly spinning neutron stars that, in their way, are as exciting to astronomers as little green men would be. The pulsar PSR 1937+21 is ticking away with greater precision than the best atomic clocks. It beams out a pulse of radio energy every 1.55780644887275 thousandths of a second—plus or minus three milliseconds.

The proof of the neutron-star origin for the pulsars came from the Crab Nebula. This is a wildly distorted batch of glowing gas in the constellation of Taurus, the Bull. You can see it for yourself with a moderate-sized telescope. In the year A.D. 1054 Chinese and Japanese astronomers recorded a "new star" that was so bright when it first appeared that it could be seen in broad daylight. We now recognize that it was a titanic stellar explosion. The Crab Nebula is the remains of that shattered star. And at the heart of the Crab Nebula is a pulsar. It not only sends out bursts of radio energy, it also winks on and off in visible wavelengths. Its picture made the front page of the *New York Times* when it was discovered in the late 1960s.

The "new star" of 1054 was what astronomers today call a *supernova. Nova* is the Latin word for "new," and from time to

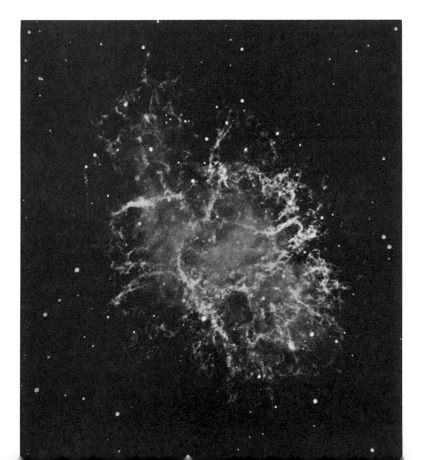

The Crab Nebula, remnants of a star that tore itself apart in a supernova explosion that was seen on Earth in A.D. 1054. At the heart of the nebula there is a tiny pulsar, all that remains of the original star.

time astronomers have seen stars appear in the heavens where no star was observable before. After a few days, or perhaps weeks, the star dims and disappears from view. Of course there was a star there all the time, but it was too faint to be seen from Earth until it erupted. Many stars go nova; some of them do it repeatedly, puffing away like stellar steam engines.

But a supernova is different. Supernova explosions are the most violent kind of explosion that a star can undergo. No star can go through a supernova explosion more than once.

A supernova appeared in 1572, brighter than the planet Venus, and turned Tycho from a career in the law to astronomy. Kepler also saw a supernova, in 1604, despite his poor eyesight; it was as bright as Jupiter.

It was one of those ironies of history that two brilliant supernovas exploded just before the invention of the telescope, but no supernova near enough to be seen with the naked eye flared up afterward.

Until 1987.

Astronomers observed far-distant supernovas in galaxies millions of light-years away, but they yearned for one that would be close enough to study with the full panoply of their growing storehouse of optical, electronic, and other instruments.

At about 2 A.M. on February 24, 1987, at the 8,000-foot-altitude Las Campanas Observatory in northern Chile, Canadian astronomer Ian Shelton noticed a new star on the photographic plate he was developing after a night's observations. At first he thought it was a flaw in the plate. Then he walked out into the windswept darkness and saw it with his unaided eyes. (See Plate 26 following page 238.)

It was named Supernova 1987a, after the astronomical community's system of nomenclature. The supernova is in the Large Magellanic Cloud, a near companion of our Milky Way galaxy, barely 170,000 light-years distant.

Within hours the world's astronomers and astrophysicists were training their instruments and attention to Supernova 1987a, excited beyond measure by this fantastic opportunity. It was the chance of a lifetime, a supernova close enough to scrutinize carefully. Imagine your own excitement if you had heard all your life about some fabulous event, and suddenly it was *there*, practically in your own backyard!

Not merely visible light, but X-rays, radio waves, even ghostly subatomic particles called neutrinos have been observed from Supernova 1987a. Radiation from the element cobalt-56, one of the elements that can be formed only in a supernova explosion if astrophysical theory is correct, has been detected. In fact, the supernova's light faded during the summer of 1987 at exactly the rate that matches the radioactive decay of cobalt-56 into iron-56. It will take years before all the data are carefully analyzed; the supernova is still blazing in the sky as I write these words.

The conclusions reached by the scientists so far, based on their preliminary observations of 1987a, show that the earlier theories developed about supernovas are standing up rather well in the light of the new information. Supernova 1987a is not behaving precisely as the theories said it should, but that is to be expected. After all, no one has seen a supernova this close before. The star is a classroom, and the world's astronomers and physicists are attending eagerly.

In the November 13, 1987, issue of *Science* magazine, astrophysicists from the California Institute of Technology raised an intriguing question: Will we see a pulsar at the heart of Supernova 1987a? If we do, it is powerful confirmation of our basic understanding of stellar astrophysics. The Caltech scientists suggest looking for gamma rays coming from the supernova as the earliest indication that a pulsar has formed at its heart.

The Sun will not become a supernova. It is unlikely that the Sun will even go nova. Instead of exploding, the Sun will merely collapse down to a white dwarf—after bloating enough to boil life off the planet Earth.

(A personal aside. The true purpose of space flight is to ensure the human race's survival. Long before the Sun begins to threaten life on Earth, natural disasters or human folly could make our planet uninhabitable. Space flight can lead to self-sufficient human habitats elsewhere. In essence, space flight allows us to detach the fate of the human race from the fate of the Earth. *Star* flight would allow us to detach the fate of the human race from the fate of the Sun.)

Remember that throughout a star's lifetime, it is subject to two forces that exert themselves in opposite directions: gravity, pulling inward, and gas and radiation pressure, pushing outward. When the Sun collapses into white dwarfdom, it will be the interior pressure of its very dense matter that stops the gravitational collapse when the Sun has reached a diameter somewhat similar to Earth's. For more massive stars that undergo supernova explosions, gravity pulls harder. Their cores go beyond the white dwarf state and become neutron stars with densities of billions of tons per cubic inch. The rigidity of the neutronic matter finally balances gravity, and the neutron star's collapse is stopped when it reaches a diameter of a few miles.

For even more massive stars, however, the collapse does not stop at all. Gravity wins. Its inward-pulling force overwhelms all other forces, and the star literally disappears from the universe. It collapses into a *black hole,* where the billions upon billions of tons of the star's matter shrink down to a dimensionless point. The gravitational force becomes so strong that not even photons of light can struggle away from it, and the star disappears entirely.

Artist's concept of a black hole. When a massive star collapses, it can literally disappear from our universe.

Not even light can escape from the monstrous gravitational force of a black hole. The star winks out forever.

It's uncanny. Picture a supergiant star collapsing, smaller, smaller, until there is nothing left at all except an invisible "pothole" of gravitational energy waiting to lure anything that gets close enough into its one-way trip to oblivion. To paraphrase an old saw, the star digs a hole, jumps in, and then pulls the hole in after itself.

No one can observe black holes. By definition. They may not even exist, although astrophysical theory shows that they should. Astronomical satellites have detected strong X-ray emissions from discrete spots in the sky. Astrophysicists have calculated that powerful X-rays and gamma rays would be emitted from a star being squeezed into the ultimate collapse of a black hole.

Realize, though, that as a star is dying it spews out much of its matter into interstellar space. Most of the elements that the star has built up over the eons of its existence are blasted out into space. In a supernova explosion nuclear reactions take place that produce elements heavier than iron. The spectra of distant supernovas have even suggested that very short-lived elements such as californium, which exist on Earth only when they are briefly produced in nuclear physics laboratories, are manufactured in the hellish fury of a supernova.

All those elements waft through space and become the building material for new stars. The elements in the Sun that are heavier than hydrogen were cooked up in other, older stars that exploded eons ago. The atoms in your body of carbon, nitrogen, zinc, iron—they all came from long-dead stars.

We are stardust. Quite literally, we are made of stardust. The atoms of our bodies and of our Earth were created inside ancient stars long eons ago.

As astronomers were puzzling out the life cycles of the stars, they were also trying to map the heavens. They found a curious thing, a thing that somewhat frightened them.

The Sun appeared to be in the center of the universe.

Once Galileo turned his telescope to the sky it became abundantly clear that the 2,000 or so stars we can see with the unaided eye are only a paltry fraction of the myriad upon myriad stars that actually exist in the heavens. The Milky Way, that glowing sheen of light that wends across the sky, turned out to be made of stars, stars, and even more stars.

Every increase in telescopic power showed still more stars, clouds and swarms of stars, clusters and globules of them. So many that they were literally beyond counting. Astronomers saw other things in space, too: great clouds of shining gas, patches and globules of dark obscuring dust, and fuzzy nebulas that appeared to be spindle shaped and elliptical. (*Nebula,* meaning "cloud," is a catchall term astronomers tend to use for anything that appears faint and fuzzy.)

How far did the Milky Way extend? And where is the Sun positioned within it? To find the answers to these questions the astronomers conducted star counts. They could not hope to count each and every individual star in the heavens, so they selected a number of small sections of the sky, spread evenly in all directions, and counted the stars in each section. They reasoned that this sampling method would give as good a picture of the distribution of stars across the sky as counting each individual star.

The answer they got troubled them.

The Milky Way, it seemed, was a flat disc of stars, shaped rather like a saucer. And the Sun was smack in its center.

They counted again. And again. By the beginning of the twentieth century the best available evidence showed that the Sun was very close to the center of the Milky Way.

Most astronomers found this conclusion difficult to accept. After all, the lesson of Copernicus is that this is not a geocentric universe. The Earth is not the center of the universe. Why should the Sun be? It is a rather ordinary G-class yellow star, average in every discernible way. Why should it be the center of creation?

The evidence was strong, though; the conclusion inescapable. And wrong.

The American astronomer Harlow Shapley finally showed why the conclusion was wrong and the evidence misleading. He noticed that there were a hundred or so great globular star clusters—spherical clusters that contain hundreds of thousands of stars—all bunched in one sector of the sky. With a faith that nature tends to be symmetrical, Shapley concluded that these clusters are actually grouped around the true center of the Milky Way, and they appear in one section of our sky because we are far from that center.

Shapley's argument was based largely on his faith in symmetry. Within a few years he was shown to be right. The globular clusters *are* arranged around the Milky Way's center, and our Solar System lies some 30,000 light-years from the Milky Way's hub.

Then what about the star counts? It turns out that nature was pulling a little trick on the astronomers. Interstellar space is not crystal clear. Astronomers had seen and photographed many large formations of dark, dust-laden nebulas. What they did not realize was that there are lanes of dust within the Milky Way that dim the light of the stars and obscure entire sections of it from our view. The blazing heart of the Milky Way is totally hidden from our sight by thick clouds of obscuring dust.

Shapley could see the globular clusters because they hover high above (and below, in the Southern Hemisphere) these dust lanes.

A famous debate took place in Washington, D.C., at a meeting of the National Academy of Science on April 26, 1920. Shapley argued against Heber D. Curtis of the Lick Observatory. The debate considered two points: (1) Is the Sun at the center of the Milky Way? and (2) Does the Milky Way encompass the entire universe?

Shapley argued that the Sun is not at the center of the Milky Way and that the Milky Way does indeed make up the whole universe. He was right about the Sun, wrong about the Milky Way.

The argument was not settled that night. It took more years of observation and measurement.

The 100-inch reflector of the Mount Wilson observatory went into operation in 1917. (And ceased operation, sadly, in 1985, a victim of the pollution caused by the relentless growth of Los Angeles.) By the time of the Shapley-Curtis debate, the Mount Wilson telescope was producing photographs of those faint spindle- and elliptical-shaped nebulas that showed they contained stars in them. They were not clouds at all but islands of stars floating millions of light-years distant from the Milky Way.

Today we call them *galaxies*. This is a term much abused in Hollywood's science-fiction productions. You would think that, being so close to the greatest astronomical observatories in the world, they could get the wording right. But they do not even bother to try. If ignorance is bliss, Hollywood is happiness sublime. In one sci-fi epic after another, the word *galaxy* is used to

The approximate position of our Solar System in the Milky Way galaxy. It takes 200 million years for the Solar System to complete one orbit around the galaxy's center.

mean a planetary system, like our own Solar System—a star that has planets orbiting around it.

A galaxy, as astronomers use the term, is a giant colony of stars. Our Milky Way is a galaxy that contains something like 100 billion stars, plus enough loose gas and dust to make billions more. The great spiral "nebula" in the constellation Andromeda is actually a spiral galaxy very much like our Milky Way. Astronomers say it is "in Andromeda" as a key to where to look for it: You find the constellation Andromeda and look through it to see the galaxy, some 2 million light-years away. It is like saying that you can see the neighbors across the street by looking through the living room window.

Far from being the "all" of the universe, our Milky Way is one galaxy among billions of galaxies. It is one of the larger galaxies, spiral in shape, measuring roughly 100,000 light-years across and about 20,000 light-years at its thickest, the central bulge. Our Solar System is some 30,000 light-years from its center, wheeling around in an orbit that takes 200 million years to complete.

Galaxies are composed of stars. The universe is composed of galaxies. Galaxies cluster together too. Our own Milky Way is part of what astronomers call the Local Group (not the most romantic of appellations), a little cluster of 16 galaxies, the largest of which are the Milky Way itself and the great spiral in Andromeda, which appears to be a near-twin of the Milky Way. Other clusters of galaxies, deeper in space, have hundreds of members, even thousands. And the clusters seem to be lumped together in superclusters.

In 1929 the New York Stock Exchange crashed and the universe exploded. The two events had nothing to do with each other. Actually, the universe had apparently exploded a lot earlier. We only found out about it in '29.

In that year the American astronomer Edwin P. Hubble (for whom the Hubble Space Telescope has been named) announced that most of the galaxies in the universe appear to be rushing away

The observable universe contains billions of galaxies. Many are spiral in shape; others are elliptical, and a few have irregular shapes.

from us. Citing observations that the spectra of the galaxies are shifted toward the red end of the spectrum, Hubble and others interpreted this shift as a movement away from us.

The astronomers saw an apparent Doppler shift in the spectra of the distant galaxies. This redshift startled the world's cosmologists much more than the stock market crash. Cosmologists are astronomers, physicists, and others who study the universe itself rather than its parts. An astronomer can be a cosmologist merely by changing emphasis. A particle physicist probing the inner workings of the atomic nucleus can be a cosmologist when he applies his knowledge to the workings of the universe as a whole. Cosmology is a grand game, particularly when you start to consider the question of how the universe began.

The big problem with cosmology is that we are inside the universe and cannot see it in its entirety. That is why Hubble's redshift announcement was so important. It was one of those precious few observations that took in the entire universe. All of it.

There is a cosmic irony here. Galactic spectra are photographed by the biggest and most powerful telescopes ever built, and then the photographic film is inspected minutely in microscopes. To determine the redshift of a galaxy, astronomers peer into microscopes and compare the galaxy's spectrum to a standard laboratory source, measuring the percentage of displacement of key absorption or emission lines, such as the lines representing hydrogen.

The conclusions that the cosmologists reached are these:

The redshifts are caused by the Doppler effect; they represent motion of the galaxies away from us. The 16 nearest galaxies of our Local Group appear to be gravitationally bound to one another and are not flying away from each other. But galaxies beyond the Local Group are all redshifted. Other clusters of galaxies are moving away as groups. Apparently gravity is a strong-enough force to hold clusters of galaxies together even though the clusters are rushing away from one another. Whatever is causing the redshift movement operates at longer range.

The farther a galaxy is, the faster it appears to be receding. This is a key observation. It means that the redshift of a very distant galaxy can be used as a measurement of its distance, where other distance-measurement techniques do not work.

RELATION BETWEEN RED-SHIFT AND DISTANCE
FOR EXTRAGALACTIC NEBULAE

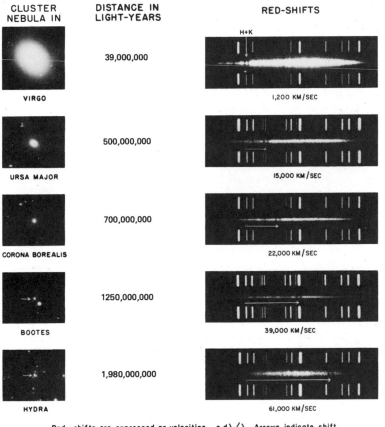

CLUSTER NEBULA IN	DISTANCE IN LIGHT-YEARS	RED-SHIFTS
VIRGO	39,000,000	1,200 KM/SEC
URSA MAJOR	500,000,000	15,000 KM/SEC
CORONA BOREALIS	700,000,000	22,000 KM/SEC
BOOTES	1250,000,000	39,000 KM/SEC
HYDRA	1,980,000,000	61,000 KM/SEC

Red-shifts are expressed as velocities, $c\,d\lambda/\lambda$. Arrows indicate shift for calcium lines H and K. One light-year equals about 9.5 trillion kilometers, or 9.5×10^{12} kilometers.

Distances are based on an expansion rate of 50 km/sec per million parsecs.

The galaxies beyond our own Local Group all show redshifted spectra, which indicates they are moving away from us. Cosmologists interpret this to mean that the universe is expanding.

The galaxies are not merely rushing away from us. We are not the center of the universe, and we are not afflicted with some sort of cosmic halitosis. The very fabric of the universe itself is expanding, and all the galaxies (or galactic clusters) are moving away from all the others. It merely looks as if they are moving away from us because of our point of view. If we could be transported to a distant galaxy, it would look as if everything is rushing away from it.

Cosmologists use an analogy here to help picture the situation. Imagine a balloon with tiny dots painted on its surface to represent the galaxies. As you inflate the balloon, the dots all move farther away from one another. That is what is happening to the universe. It is expanding.

Why?

The simplest answer is that the universe began in an incredible explosion some 15 or 20 billion years ago, and the galaxies are still being flung apart as its result. The galaxies are celestial shrapnel, according to this Big Bang cosmological theory.

The Steady State theory of the late 1940s claimed otherwise. According to Bondi, Gold, and Hoyle the universe had no beginning. It always was and always will be. It is expanding because new matter is being created, perhaps one single hydrogen atom at a time, out of nothingness. The new matter entering the universe forces the rest of the universe to expand to make room for it.

There are some beautiful aspects to the Steady State theory, but most cosmologists have rejected it. For two reasons.

The first reason is the quasars.

In the 1960s radio astronomers found very powerful radio sources in the heavens that appeared to be coming from bluish stars. They called the sources *quasi-stellar objects;* "quasi-stellar" because, while these things sort of looked like stars, they had some very definite peculiarities about them. For one thing, their spectra were wonky, unlike those of any star seen before. It turned out that the spectra are redshifted, enormously redshifted. The quasar redshifts indicate that they are far outside the Milky Way—many of them much farther away than any known galaxy! (See Plate 27 following page 238.)

And they are emitting prodigious amounts of energy, more than a thousand normal galaxies combined.

And they are pulsating. Both in visible and radio wavelengths, they grow brighter and dimmer, often in a matter of mere days. That means that they have to be no more than a few light-days in diameter, since the brightening and dimming could not possibly move across the quasar faster than the speed of light. Could it?

Objects only slightly larger than our own Solar System, emitting more energy than a thousand Milky Ways, receding from us

so fast that they must be out at the very edge of the observable universe.

The quasars presented an incredible puzzle to cosmologists, astronomers, astrophysicists—even to science-fiction writers.

Some thinkers came to the conclusion that the quasar redshifts have nothing to do with distance. If the quasars are not "cosmologically" distant, but are "local" objects, close to the Milky Way or even within it, then the estimates of their energy production can be scaled down. Instead of being fantastic and mind-boggling, they become merely mysterious and puzzling.

But if the quasar redshifts are not related to their distances, what does that say about the whole redshift story and the assumption that the universe is expanding? Most astronomers refuse to accept the idea that the very foundation of current cosmological thought is wrong; they accept the quasars as cosmologically distant—more than 10 billion light-years away.*

Recognize that as we peer at objects that are farther and farther away, we are looking backward in time. When we look at the Moon we see it as it was about 1.3 seconds ago; it takes that long for its light to reach us. We see the Sun as it was eight minutes ago, the spiral galaxy in Andromeda as it was 2 million years ago, the most distant quasars as they were more than 10 billion years ago.

Most astronomers believe that the quasars are showing us a picture of the very early days of the universe, when titanic forces were at work in the hearts of galaxies. They postulate that the quasars are very young galaxies, with something very powerful blazing away at their cores; perhaps gigantic black holes are at the heart of each quasar, sucking in stars by the thousands and spewing out incredible amounts of energy in the process.

But there are other possibilities.

Early in 1988 astronomers from the California Institute of

---

*The farthest objects yet detected have redshifts of more than four. That is, the emission/absorption lines in their spectra are shifted more than 400 percent from where the lines would be if the light source was at rest. This corresponds to a speed of recession of more than 90 percent of the velocity of light and a distance of more than 10 billion light-years—if the redshifts are truly Doppler shifts and directly linked to distance.

Technology announced that an orbiting satellite had observed what may be the birth of a quasar. The Infrared Astronomy Satellite (IRAS), launched into orbit in 1981, has detected galaxies that are extremely luminous in infrared radiation. One such object, IRAS 14348-1447, is more than 100 times brighter than our Milky Way galaxy, with more than 95 percent of its energy in the far-infrared region of the spectrum. It appears to be a pair of galaxies that are colliding, crashing into one another. Astronomers D. S. Sanders, N. Z. Scoville, and B. T. Soifer believe that 14348-1447 is a colliding pair of galaxies enshrouded by thick clouds of dust and gas, more than 20 times the amount of interstellar dust and gas in the Milky Way. They suggest that quasars are "born" when such galaxies collide with one another.

Older, more sedate galaxies also have active central regions, though not as active as the quasars' cores. Our own Milky Way galaxy may have a gigantic black hole chewing at its core, gobbling whole stars and emitting radio and high-energy X- and gamma rays. While we cannot see the Milky Way's heart in optical wavelengths, because of the dust obscuring it, radio and infrared wavelengths penetrate the dust and reveal that *something* very energetic is happening at the heart of our galaxy.

If the Big Bang theory is correct, then the universe is changing over time, evolving. Unlike the placid sameness of the Steady State theory, the Big Bang begins with a cosmic explosion that starts the expansion of the universe. If this is so, then the bits and pieces of the universe must have been clustered closer together in the earlier epochs of time. The galaxies that we see today are spreading apart, if their redshifts are to be believed. As we look further and further back in time, we should see galaxies huddled in denser formations.

If the quasars are cosmologically distant they are the farthest objects we can see. Therefore they are the youngest; we are seeing them as they existed some 10 billion years ago or more. Thus they may be very young galaxies. If they are grouped more densely than nearer, older galaxies, it would indicate that the galaxies were closer together far back in the past. This would be powerful evidence in favor of the Big Bang theory. But the observations made so far on the 5,000-some quasars that have been identified have not been able to positively determine whether the quasars are

packed more densely than the galaxies closer to us in space and time.

Still, most astronomers and cosmologists accept the Big Bang theory as the best available description of the observable universe and its origin.

In this view, the universe began as a sort of cosmic egg, with all the matter and energy that we now observe squeezed together in one tiny package.

It exploded. The cosmic fireball, with temperatures so high that the numbers become meaningless, expanded—creating space where nothing existed before. At first only space and energy existed. Particles of matter and the forces that we know as electromagnetism, gravity, and the nuclear forces took form out of the chaos of that primeval explosion. As the fireball continued to expand and cool, protons and electrons formed hydrogen atoms and then helium atoms.

That was enough to begin the creation of the stars. All the elements heavier than helium were cooked in the nuclear ovens at the cores of the stars. They are still being created now.

The expansion of the universe that is inferred by the redshifts of the galaxies and quasars is the result of that original primeval Big Bang. Will the expansion go on forever, the galaxies drifting ever farther apart? Or will the expansion ultimately slow, even reverse itself, into a stupendous collapse that ends the universe in a new cosmic fireball? That depends on the total amount of matter in the universe; is there enough to eventually put gravitational brakes on the expansion? No one knows. Astrophysicists suspect that most of the universe's matter is dark and invisible to us; the bright stars and gas clouds we see may represent only a small fraction of the total matter. But no experiment or observation yet conceived gives promise of detecting the universe's dark matter—if it exists.

And some of the bright matter astronomers have photographed may not exist! One of the consequences of Einstein's relativity is that massive galaxies may act as "gravitational lenses," bending light with their powerful gravitational fields much as a lens or mirror would. Some photos of quasars and vast arcs of luminous gas stretching between galaxies have been shown to be optical illusions created by gravitational lenses deep in intergalactic space.

If the Big Bang theory is a correct picture of the universe, there should be some evidence of the original explosion. Out in deepest space there ought to be some remnant of that titanic event that we can observe and measure.

And there is.

In 1964 Arno A. Penzias and Robert W. Wilson, working at the Bell Telephone Laboratories, detected faint radio-wavelength radiation coming from all points of the sky. With the help of Princeton physicist Robert H. Dicke they determined that the radio waves are a residue of the original Big Bang—so cooled over the billions of years since its explosion that the radiation is now only three degrees above absolute zero.

This faint radiation is assumed by most cosmologists to be the afterglow of the explosion that created the universe, the pale ghostly shadow of the cosmic explosion that started it all, the last remnant of that titanic event that began, according to Genesis, with the words:

"LET THERE BE LIGHT!"

Bibliography

Alpern, Mathew. "Color Blind Color Vision." *Trends in NeuroSciences* 4, no. 6 (June 1981).

Barnes, R.D. *Invertebrate Zoology*. Philadelphia: Saunders College, 1980.

Beatty, J. Kelly, Brian O'Leary, and Andrew Chaikin. *The New Solar System*. Cambridge, MA: Sky Publishing Corp., 1981.

Bova, Ben. *The New Astronomies*. New York: St. Martin's Press, 1972.

————. *In Quest of Quasars*. New York: Crowell-Collier Press, 1969.

Bowmaker, J.K. "Trichromatic Colour Vision: Why Only Three Receptor Channels?" *Trends in NeuroSciences* 6, no. 2 (February 1983).

Boyer, Carl B. *The Rainbow*. Princeton, NJ: Princeton University Press, 1987.

Branley, Franklin M., and Mark R. Chartrand III. *Astronomy*. New York: Thomas Y. Crowell Co., 1975.

Bronowski, Jacob. *The Ascent of Man*. Boston: Little, Brown and Company, 1973.

Brou, Philippe, et al. "The Colors of Things." *Scientific American* 255, no. 3 (September 1986).

Campbell, Jeremy. *Winston Churchill's Afternoon Nap*. New York: Simon and Schuster, 1986.

Collis, John Stewart. *The World of Light*. New York: Horizon Press, 1960.

Coon, Carleton S. *The Story of Man*. 3d ed., rev., New York: Alfred A. Knopf, 1971.

Dubkin, Louis R. *Factbook of Man From Birth to Death*. 2d ed. New York: The Macmillan Co., 1969.

Dubos, Rene. *The Wooing of Earth*. New York: Charles Scribner's Sons, 1980.

Finke, Ronald. "Mental Imagery and the Visual System." *Scientific American* 254, no. 3 (March 1986).

Forester, Tom. *High-Tech Society*. Cambridge, MA: The MIT Press, 1987.

Gallant, Ray. *Our Universe*. Washington D.C.: National Geographic Society, 1980.

Glashow, Sheldon, and Ben Bova. *Interactions*. New York: Warner Books, 1988.

Gombrich, E.H. *The Story of Art*. 12th ed. London: Phaidon, 1972.

Grimal, Pierre. *The Dictionary of Classical Mythology*. Oxford: Basil Blackwell Publisher, 1986.

Hartmann, William K. *Moons and Planets: An Introduction to Planetary Science*. Belmont, CA: Wadsworth Publishing Co., Inc., 1973.

Hecht, Jeff, and Dick Teresi. *Laser*. New Haven and New York: Ticknor & Fields, 1982.

Keeton, W.T. *Biological Science*. New York: W.W. Norton & Co., 1972.

Kirschfeld, K. "Are Photoreceptors Optimal?" *Trends in NeuroSciences* 6, no. 3 (March 1983).

Klein, Miles V., and Thomas E. Furtak. *Optics*. New York: John Wiley & Sons, 1986.

Kluge, P.F. "A Diamond as Big as the Ritz—Well, Just About as Big." *Smithsonian,* May 1988.

Landsberg, H.E. *Weather and Health*. New York: Anchor Books, Doubleday & Co., 1969.

Livingstone, Margaret, and David Hubel, "Segregation of Form, Color, Movement, and Depth: Anatomy, Physiology, and Perception." *Science* 240, no. 4853 (6 May 1988).

Masland, Richard H. "The Functional Architecture of the Retina." *Scientific American* 255, no. 6 (December 1986).

Mayo, John S. "Materials for Information and Communication." *Scientific American* 255, no. 4 (October 1986).

Motz, Lloyd, ed. *Rediscovery of the Earth*. New York: Van Nostrand and Reinhold Co., 1979.

Nassau, Kurt. *The Physics and Chemistry of Color*. New York: John Wiley & Sons, 1983.

Ogburn, Charlton. *The Marauders*. New York: Harper & Row, 1959.

Ornstein, Robert, and Richard F. Thompson. *The Amazing Brain*. Boston: Houghton Mifflin Co., 1984.

Palmer, Richard H. *The Lighting Art*. Englewood Cliffs, NJ: Prentice-Hall, Inc., 1985.

Pilbrow, Richard. *Stage Lighting*. New York: Drama Book Specialists, 1979.

Pryensu, Edward S., and Philip Whitfield. *The Rhythms of Life*. New York: Crown Publishers Inc., 1981.

Ramachandran, Vilayanur, and Stuart M. Antsis. "The Perception of Apparent Motion." *Scientific American* 254, no. 6 (June 1986).

Rhea, John. "Fly by Light." *Air Force Magazine,* March 1988.

Rose, Kenneth Jon. *The Body in Time*. New York: John Wiley & Sons, 1988.

Rowell, J.M. "Photonic Materials." *Scientific American* 255, no. 4 (October 1986).

Schnapf, Julie L., and Denis A. Baylor. "How Photoreceptor Cells Respond to Light." *Scientific American* 256, no. 4 (April 1987).

Schneer, Cecil J. *The Search for Order*. New York: Harper & Brothers, 1960.

Schneider, Stephen H., and Randi Londer. *The Coevolution of Climate and Life*. San Francisco: Sierra Club Books, 1984.

Sellman, Hunton D., and Merrill Lessley. *Essentials of Stage Lighting*. Englewood Cliffs, NJ: Prentice-Hall, Inc., 1982.

Simpson, G.G., and W.S. Beck. *Life: An Introduction to Biology*. New York: Harcourt, Brace and World Inc., 1965.

Sobel, Michael I. *Light*. Chicago: The University of Chicago Press, 1987.

Speer, Albert. *Inside the Third Reich*. New York: Macmillan Publishing Co., 1970.

Stryer, Lubert. "The Molecules of Visual Excitation." *Scientific American* 257, no. 1 (July 1987).

Stuart, Anne E. "Vision in Barnacles." *Trends in NeuroScience* 6, no. 4 (April 1983).

Treisman, Anne. "Features and Objects in Visual Processing." *Scientific American* 255, no. 5 (November 1986).

Williamson, Samuel J., and Herman Z. Cummins. *Light and Color in Nature and Art*. New York: John Wiley and Sons, 1983.

Wurtman, Richard J., et al., eds. *The Medical and Biological Effects of Light*. New York: New York Academy of Sciences, 1985.

Index